Investigation of the Helios Prototype Aircraft Mishap

Volume I
Mishap Report

Thomas E. Noll
NASA Langley Research Center

John M. Brown
National Oceanic and Atmospheric Administration

Marla E. Perez-Davis
NASA Glenn Research Center

Stephen D. Ishmael
NASA Dryden Flight Research Center

Geary C. Tiffany
NASA Ames Research Center

and

Matthew Gaier
NASA Headquarters

January 2004

Table of Contents

Volume I – Mishap Report

Table of Contents (continued)

Acknowledgements

The Helios Mishap Investigation Board wishes to thank the Independent Working Group Teams from Dryden Flight Research Center (Marty Brenner, Bob Clarke, and Chan-gi Pak), Glenn Research Center (Anastacio Baez, Mark Kankam, and Thomas Miller), and Langley Research Center (Damodar Ambur, Gary Farley, David McGowan, Danniella Muheim, Anthony Pototzky, Rob Scott, Walt Silva, David Sleight, and John Wang) for their cooperation, spirited discussions, and excellent written and oral reports. Also, the Project personnel at the Dryden Flight Research Center, namely John Del Frate, Robert Navarro, Mauricio Rivas, and Lesa Brown, were always available to assist the Board in their needs.

The Board wishes to thank Robert Sharman of the National Center for Atmospheric Research, Duane Stevens and John Porter of the University of Hawaii, and Jack Ehrenberger of the Dryden Flight Research Center for their assistance in understanding the weather environment around the island of Kauai. Also, the procurement and contracting personnel at the Kennedy Space Center are greatly appreciated for their timely assistance in awarding contracts.

The Board wishes to acknowledge the Ames Research Center for making their Investigative Organizer Tool available for use in the investigation. A very special thanks goes to Anita Abrego and Theodore Forsyth for their hard work in instructing the Board, inputting data into the system, and supporting the Board members in using the tool.

Finally, the Board wishes to thank certain individuals at AeroVironment for their invaluable contributions. We thank the technical personnel for assisting the Board in understanding the evolution of the Helios Prototype aircraft, the events leading up to the mishap, and the complex interactions that caused the mishap. In particular, the Board wishes to acknowledge Derek Lisoski, Bart Hibbs, Dana Taylor, and Bill Parks. Also, the Board thanks Lana Danta for outstanding administrative support provided to the Board.

Section 1
Transmittal Letter

National Aeronautics and
Space Administration

Langley Research Center
Hampton, VA 23681-2199

Reply to Attn of: 121

January 16, 2004

TO: NASA Headquarters
           Attn: R/Associate Administrator, Office of Aeronautics

FROM: 121/Chairperson, Helios Prototype Aircraft Mishap

SUBJECT: Investigation Report of the Helios Prototype Aircraft Mishap

Reference your letter dated June 27, 2003, which established the Helios Prototype Aircraft
Mishap Investigation Board and defined the Board's responsibilities.

The Mishap Investigation Board has completed its investigation. This memorandum transmits
the written report of the Board's activities and findings.

Thomas E. Noll, Ph.D., P.E.

3 Enclosures:
Hard Copy of Written Report
CD of Written Report
DVD of Mishap

Section 2
Signature Page

Concurrence by Board Members:

Thomas E Noll   12 Jan 04

Chairperson, Thomas E. Noll
Deputy Director, Structures and Materials Competency, NASA Langley Research Center

John M. Brown   23 JAN 04

John M. Brown
Meteorologist, National Oceanic and Atmospheric Administration, Forecast Systems Laboratory

signature   20 JAN 04

Marla E. Perez-Davis
Chief, Electrochemistry Branch, NASA Glenn Research Center

signature   27 JAN 04

Stephen D. Ishmael
Special Assistant to the Director, NASA Dryden Flight Research Center

signature   4 Feb 04

Geary C. Tiffany
Chief, Aviation Management Office, NASA Ames Research Center

signature   14 JAN 04

Matthew Gaier
Ex-officio, Office of Safety & Mission Assurance, NASA Headquarters

6

Section 3
List of Members, Advisors, Observers, and Others

The Associate Administrator, Office of Aeronautics, National Aeronautics and Space Administration (NASA) established the Board on 27 June 2003, to investigate the Helios prototype aircraft mishap. The Memorandum of Appointment is included as Document B.1 in Appendix B, Volume II. The Mishap Investigation Board (MIB) was identified in the Memorandum of Appointment, however to be consistent with NASA policy defined in NPG 8621.1, a letter (Document B.2 in Appendix B, Volume II) dated 14 August 2003, was prepared by the Board Chairperson that specifically identified the Ex officio member to the Board and the advisors to the Board. With these changes, the MIB consisted of the following individuals with the assigned responsibilities:

Chairperson:
(voting)
Dr. Thomas E. Noll, Deputy Director, Structures and Materials Competency, Langley Research Center (LaRC)

Members:
(voting)
Dr. John M. Brown, Meteorologist, National Oceanic and Atmospheric Administration (NOAA), Forecast Systems Laboratory

Dr. Marla E. Perez-Davis, Chief, Electrochemistry Branch, Glenn Research Center (GRC)

Mr. Stephen D. Ishmael, Special Assistant to the Director, Dryden Flight Research Center (DFRC)

Mr. Geary C. Tiffany, Chief, Aviation Management Office, Ames Research Center (ARC)

Ex-officio
Member:
(nonvoting)
Mr. Matthew Gaier, NASA Aviation Safety Manager, Office of Safety & Mission Assurance, NASA Headquarters

Advisors:
(nonvoting)
Mr. Bart Henwood, Aviation Safety Manager, DFRC
Mr. John Madura, Weather Office, Kennedy Space Center (KSC)
Mr. Ted Wierzbanowski, Managing Director, AeroVironment, Inc. (AV)

Supporting
Staff:
(nonvoting)
Mr. Robert Anderson, Office of Aeronautics, NASA Headquarters
Mr. Gary Krier, Flight Operations, DFRC
Mr. Vance Brand, Aerospace Projects, DFRC
Mr. John Sharkey, Aerospace Projects, DFRC
Mr. Larry Crawford, Research Engineering, DFRC
Mr. Lawrence Davis, Office of Safety and Mission Assurance, DFRC
Ms. Jennifer Baer-Reidhart, Public Affairs, DFRC
Mr. David Samuels, Chief Counsel, DFRC

The Memorandum of Appointment authorized the Board to:

(1) Obtain and analyze whatever evidence, facts, and opinions it considers relevant. The Board will use reports of studies, findings, recommendations, and other actions by NASA officials and contractors. The Board may conduct inquiries, hearings, tests, and other actions it deems appropriate. The Board may take and receive privileged statements from witnesses.

(2) Impound property, equipment, and records as necessary.

(3) Determine the actual cause(s) or if unable, determine probable cause(s) of the Helios Prototype Aircraft Mishap, and document and prioritize their findings in terms of (a) proximate cause(s) of the mishap, (b) root cause(s), and (c) contributing factor(s), and (d) significant observation(s).

(4) Develop recommendations for preventive or other appropriate actions.

(5) Provide a verbal report to the Associate Administrator, Office of Aeronautics, as soon as possible, and a final report in the format specified in NPG 8621.1.

(6) Provide a lessons learned summary.

(7) Perform any other duties that may be requested by the Associate Administrator, Office of Aeronautics.

Section 4
Executive Summary

The Helios Prototype vehicle was one of several remotely piloted aircraft funded and developed by NASA under the Environmental Research Aircraft and Sensor Technology (ERAST) project, and managed by NASA's Dryden Flight Research Center (DFRC). This vehicle was a proof-of-concept, propeller-driven, flying wing built and operated by AeroVironment, Inc. The vehicle consisted of two configurations. One configuration, designated HP01, was designed to operate at extremely high altitudes using batteries and high-efficiency solar cells spread across the upper surface of its 247-foot wingspan. On 13 August 2001, this aircraft configuration reached an altitude of 96,863 feet, a world record for sustained horizontal flight by a winged aircraft. The other configuration, designated HP03, was designed for long-duration flight. The plan was to use the solar cells to power the vehicle's electric motors and subsystems during the day and to use a modified commercial hydrogen–air fuel cell system for use during the night. The vehicle was also equipped with batteries as a backup source of power. The aircraft design used wing dihedral, engine power, elevator control surfaces, and a stability augmentation and control system to provide aerodynamic stability and control.

On 26 June 2003, HP03-2 took off at 10:06am local time from the Navy's Pacific Missile Range Facility (PMRF) located on the island of Kauai, Hawaii. The aircraft was under the guidance of AeroVironment, Inc. (AV) ground-based mission controllers. At that time the environmental wind conditions appeared to be within an acceptable envelope, and consisted of a wind shadow over and offshore from PMRF, bounded to the north, south, and above by zones of wind shear and turbulence separating this region from the ambient easterly trade-wind flow. However, compared to previous solar-powered flights from PMRF, HP03-2 was subject to longer exposure to the low-level turbulence in the lee of Kauai due to the shallower climb out trajectory. The vehicle's longer exposure to Kauai's lee side turbulence and lower shear line penetration were superposed on what the Board now recognizes as greater airplane sensitivity to turbulence and may have been compounded by the apparent narrow corridor between the shear lines noted by the chase helicopter observer.

At 10:22am and 10:24am, the aircraft encountered turbulence and the wing dihedral became much larger than normal and mild pitch oscillations began, but quickly damped out. At about 30 minutes into the flight, the aircraft encountered turbulence and morphed into an unexpected, persistent, high dihedral configuration. As a result of the persistent high dihedral, the aircraft became unstable in a very divergent pitch mode in which the airspeed excursions from the nominal flight speed about doubled every cycle of the oscillation. The aircraft's design airspeed was subsequently exceeded and the resulting high dynamic pressures caused the wing leading edge secondary structure on the outer wing panels to fail and the solar cells and skin on the upper surface of the wing to rip off. The aircraft impacted the ocean within the confines of the PMRF test range and was destroyed. The crash caused no other property damage or any injuries to personnel on the ground. Most of the vehicle structure was recovered except the hydrogen-air fuel cell pod and two of the ten engines, which sank into the ocean.

9

The root causes of the mishap include:

- Lack of adequate analysis methods led to an inaccurate risk assessment of the effects of configuration changes leading to an inappropriate decision to fly an aircraft configuration highly sensitive to disturbances.

- Configuration changes to the aircraft, driven by programmatic and technological constraints, altered the aircraft from a spanloader to a highly point-loaded mass distribution on the same structure significantly reducing design robustness and margins of safety.

The aircraft represents a nonlinear stability and control problem involving complex interactions among the flexible structure, unsteady aerodynamics, flight control system, propulsion system, the environmental conditions, and vehicle flight dynamics. The analysis tools and solution techniques were constrained by conventional and segmented linear methodologies that did not provide the proper level of complexity to understand the technology interactions on the vehicle's stability and control characteristics. As a result, key recommendations include:

- Develop more advanced, multidisciplinary (structures, aeroelastic, aerodynamics, atmospheric, materials, propulsion, controls, etc) *"time-domain"* analysis <u>methods</u> appropriate to highly flexible, "morphing" vehicles.

- Develop ground-test procedures and techniques appropriate to this class of vehicle to validate new analysis methods and predictions.

- For highly complex projects, improve the technical insight using the expertise available from all NASA Centers.

- Develop multidisciplinary (structures, aerodynamic, controls, etc) <u>models</u>, which can describe the nonlinear dynamic behavior of aircraft modifications or perform incremental flight-testing.

- Provide adequate resources to future programs for more incremental flight-testing when large configuration changes significantly deviate from the initial design concept.

During the course of this investigation the MIB discovered that the AV/NASA technical team had created most of the world's knowledge in the area of High Altitude-Long Endurance (HALE) aircraft design, development, and test. This has placed the United States in a position of world leadership in this class of vehicle, which has significant strategic implications for the nation. The capability afforded by such vehicles is real and unique, and can enable the use of the stratosphere for many government and commercial applications. The MIB also found that this class of vehicle is orders of magnitude more complex than it appears but that the AV/NASA technical team had identified and solved the toughest technical problems. Although more knowledge can and should be pursued as recommended in this report, an adequate knowledge base now exists to design, develop, and deploy operational HALE systems.

Section 5
Method of Investigation, Board Organization,
and/or Special Circumstances

This report summarizes the combined efforts of an independent MIB investigation and AV's internal investigation. This section of the report describes the procedures used by the Board in conducting its investigation. Volume III, Appendix E provides short summaries of the AV documents referenced in this report including Document #25, which describes the details of the AV internal investigation. NASA Form 1627 (NASA Mishap Report) is included as Appendix A in Volume II.

Methods Used During MIB Investigation
The methods used by the MIB during its investigation included the following:

- Reviewing witness statements and conducting interviews (Volume IV);

- Impounding project operational and programmatic records;

- Establishing a mishap data archive;

- Receiving information/background briefings from AV and NASA personnel and from project consultants;

- Photographing and inspecting damaged components of the aircraft;

- Videotaping recovery operations;

- Viewing the operational areas of the PMRF;

- Viewing available film, video, and photos (Document C.1 in Appendix C, Volume II);

- Reviewing all existing pertinent electronic and printed information;

- Inspecting the operational facilities used by the project team including the stationary and mobile pilot stations, the data acquisition system, the telemetry data system, all weather related equipment, and areas manned by dynamics, stability and controls, and meteorological specialists, and those areas manned by the Flight Director and Mission Planning Engineer;

- Forming independent working groups (IWG) to perform analyses and tests in support of the MIB;

- Performing a root cause analysis, constructing an Events and Causal Factor Tree, identifying significant observations, and developing recommendations and lessons learned.

Investigation Organizer (IO)

Because this investigation involved Board/supporting members and investigation sites that were widely distributed geographically, and because the information being reviewed was extensive and quite diverse in terms of categories of evidence and technical disciplines, the MIB elected to use the ARC IO system, a web-based, information-sharing tool. The usefulness of the IO system to the MIB is that it combined the functionalities of a database, a document-sharing system, and a hyper-linked information navigation system. It also organized investigation information into technical and management subgroups, and it allowed browsing and downloading of documents so that pertinent information could be found and readily accessed in a timely manner. The IO system was configured to meet the distinct needs of the MIB and IWG, which allowed the MIB to store information gathered during the investigation or developed by the IWG in a systematic manner so that directions selected and tasks identified by the Board could be tracked. The development of the IO tool was facilitated by Mr. Geary Tiffany; Ms. Anita Abrego and Mr. Theodore Forsyth of ARC were instrumental in setting up and maintaining the tool.

MIB Meetings

The MIB formally met on three occasions between the time of the mishap and the submittal of the draft report to the Associate Administrator, Office of Aeronautics. The MIB was formed on 27 June 2003, the day after the mishap, and met for the first time on the Hawaiian Island of Kauai on 29 June. An interim chair was designated and on-site by the day after the mishap. For the next 10 days all members of the MIB took up a working station at the AV/NASA outpost at the PMRF to conduct their onsite investigation. After 9 days of investigation, the MIB constructed a preliminary Events and Causal Factor Tree for the mishap, developed an initial list of potential contributing causes and significant observations, developed a list of pertinent documents required for their review, developed a list of action items required of AV and NASA project members, and began planning the IWG activities.

Near the end of onsite investigation, the MIB summarized their experiences and interactions with onsite NASA, AV, and Navy personnel as follows:

- The NASA, AV, and Navy personnel provided outstanding support to the MIB.

- The personnel interviewed were very forthcoming and were eager to assist in any way to help investigate the accident.

- The NASA and AV technical leaders were proactive in the investigation and aggressive in self-criticizing their actions.

- The technical team members were highly multidisciplinary, technically strong, and very motivated and enthusiastic.

- The test team was very process/procedural oriented and followed a highly documented test methodology for guidance.

- Team synergy between the NASA and AV personnel was excellent.

- The on-site weather team was enthusiastic and well integrated into the flight crew.

During the week of 4 August 2003, the MIB met at the AV facility (August 4 through 6) in Simi Valley, California, and at the DFRC on 7 August. During this time the MIB:

- Obtained status reports of ongoing AV and IWG investigations;

- Defined objectives, user authorizations, input data categories, and input data accountability for the IO system;

- Further refined, assessed, and debated the Events and Causal Factor Tree analysis, the preliminary root and contributing causes, and the observations developed in Kauai.

At DFRC, the MIB received a briefing on NASA's Helios Recovery Plan and the HAULE Uninhabited Aerial Vehicle (UAV) Development Strategy and provided recommendations for additional long-term research in various technical disciplines which would have significant impact on the design, development, and testing of Helios type aircraft. In addition, the vehicle debris was again inspected, as there was evidence that the right wing main spar might have been damaged prior to impacting the water. As a result of this review, components of the right main wing spar from the wing tip to the section just inboard of the right wing hydrogen fuel tank were shipped to LaRC for further evaluation. The remaining aircraft debris, with the exception of the left main wing spar just inboard of the hydrogen fuel tank, all recovered motors and propellers, and the flight control computers, were released from impoundment for return to AV or for disposal.

During the week of 3 November 2003, the MIB again met in Simi Valley, California. The MIB reviewed and debated progress made by AV and the IWGs, discussed the conclusions proposed by the teams, reviewed and closed-out various paths in the Events and Causal Fault Tree analysis based on supporting documentation, and finalized the findings and recommendations for the final report.

In addition to these formal MIB meetings and IWG investigations facilitated by Board members, the Board held MIB weekly telecons and prepared weekly status reports for the Associate Administrator. There were also several meetings held at the AV facility in Simi Valley, California, by the various IWGs, and there were numerous IWG telecons each week of the investigation to insure continuous progress, alleviation/elimination of any roadblocks to progress, and coordination of activities.

Independent Working Group
To assist in performing analyses and tests required by the investigation, the MIB formed IWGs. Although the studies performed by the IWGs were initially envisioned to be separate along technical disciplines, the behavior of the Helios aircraft during the mishap indicated that the instability was most likely a highly complex interaction involving the flexible structure, the unsteady aerodynamics, the flight control system, the propulsion system, the vehicle flight dynamics, and the weather conditions.

The four IWGs formed by the Board are listed below. Also provided is a brief description of the tasks performed, the Board member(s) who facilitated each activity, and those individuals who contributed greatly to the investigation. The technical reports prepared by the IWGs are found in Appendix D in Volumes II and III; each report is summarized in Section 8 of this report.

- **Structural and Control System Modeling**
  Facilitated by Mr. Steve Ishmael (DFRC) and Mr. Ted Wierzbanowski (AV):

  This activity investigated the complex interactions that took place resulting in the aircraft achieving very high persistent wing dihedrals, a highly unstable dynamic instability, and loss of control.

  Contributions were provided by: Anthony Pototzky, Rob Scott, Walt Silva, Danniella Muheim, David Sleight, and John Wang (LaRC); Bob Clarke, Marty Brenner, and Chan-Gi Pak (DFRC); and Derek Lisoski, Dana Taylor, and Bart Hibbs (AV).

- **Power and Propulsion**
  Facilitated by Dr. Marla Perez-Davis (GRC)

  This investigation developed a complete understanding of the Helios power system, which included the motors, propellers, the onboard batteries, the solar cell system, the hydrogen-air fuel cell system, and the control system that ties these components together.

  Contributions were provided by: Anastacio Baez, Thomas Miller, and Mark Kankam (GRC).

- **Structural Integrity**
  Facilitated by Dr. Thomas Noll (LaRC)

  This study involved performing a material and structural inspection and a structural analysis to determine if the right wing main spar damage occurred prior to or during the breakup of the vehicle following departure from controlled flight, and obtaining an assessment of the main wing spar structural integrity prior to the mishap.

  Contributions were provided by: Damodar Ambur, David McGowan, and Gary Farley (LaRC).

- **Environment**
  Facilitated by Dr. John Brown (NOAA) and Mr. John Madura (KSC)

  This contracted investigation resulted in simulations and actual flight measurements of turbulence to better understand the overall flow patterns in the vicinity of Kauai, and the airflow turbulence induced by flow over and around the island.

Contributions were provided by: Bob Sharman (National Center for Atmospheric Research, NCAR), and Duane Stevens and John Porter (University of Hawaii, UH), and included the use of the Air Force Research Laboratory's (AFRL) Maui High Performance Computing Center (MHPCC) at the UH.

Section 6
ERAST and Helios Background

This section of the report provides a brief overview of the ERAST program from the Pathfinder vehicle up to the development of the Helios Prototype HP03-2, the vehicle involved in the mishap. The objective of this section of the report is to introduce the reader to the evolutionary process that the program went through and the decisions made that resulted in HP03-2 configuration. Much of the information in this section was taken from Documents #24 and #27 found in Volume III, Appendix E.

ERAST Program and the JSRA

In 1994, NASA and industry created the ERAST Alliance to further mature HALE UAV technology. The ERAST Alliance was a unique government-industry partnership that was intended to develop both a strong science capability and commercial applications for this class of vehicle. The procurement approach used by NASA to implement the program with the alliance companies, which included AV, was the Joint Sponsored Research Agreement (JSRA).

The primary objectives of the ERAST program were to develop UAV capabilities for flying at extremely high altitudes and for long durations, demonstrate payload capabilities and sensors for atmospheric research, address and resolve UAV certification and operational issues, demonstrate the UAV usefulness to scientific, government, and civil customers, and foster the emergence of a robust UAV industry in the U.S. The JSRA approach afforded a reasonable amount of flexibility during the procurement process in that the Federal Acquisition Regulations were used as guidelines rather than rules. During the vehicle development and flight operations phases, NASA safety regulations were not required to be specifically followed, allowing each party to implement their process for assuring a successful outcome. Partners were introduced to the NASA DFRC airworthiness and safety process thus providing insight into DFRC approach for reviewing the vehicle design and operational procedures. Additionally each partner submitted annual plans for government approval, specifying milestones and top-level objectives for development and test.

Although the program followed a traditional vehicle development process, the JSRA facilitated an accelerated development schedule resulting in rapid prototyping with smaller budgets to complete program milestones. Under the JSRA, NASA provided ERAST program management and oversight, flight test facilities, operational support, and cost reimbursable funding for the development efforts. AV provided project management, subsystem test facilities, aircraft development and flight operations, and cost sharing. The primary benefits to NASA were direct participation in the development of new enabling technologies, training for NASA personnel, annual flight demonstrations, access to new HALE aircraft capabilities, and positive educational outreach activities and public relations. The primary benefits to AV were a source of funding for maturing HALE technologies, training for AV personnel, and the creation of and title to aircraft developed under the program.

In 2002, NASA and AV entered into a separate JSRA that focused on furthering the development of and transitioning the HALE aircraft to other government and civil applications beyond the end of the ERAST program.

Evolution of the Spanloader Configurations to the 3 Point Mass HP03

The HP03, the aircraft involved in the mishap, was the fifth generation of all-wing aircraft designed and built by AV at its Design Development Center in Simi Valley, California, as technology demonstrators for future solar-powered high-altitude aircraft platforms for science and commercial missions. Figure 6.1 shows the relative sizes of the 5 vehicle configurations that evolved from 1994 ending with the HP03 (the long-endurance configuration).

Pathfinder (1981-1997)

Pathfinder Plus (1997-1998)

Centurion (1996-1998)

Helios Prototype (HP01), High-Altitude Configuration (1998-2002)

Helios Prototype (HP03), Long-Endurance Configuration (2003)

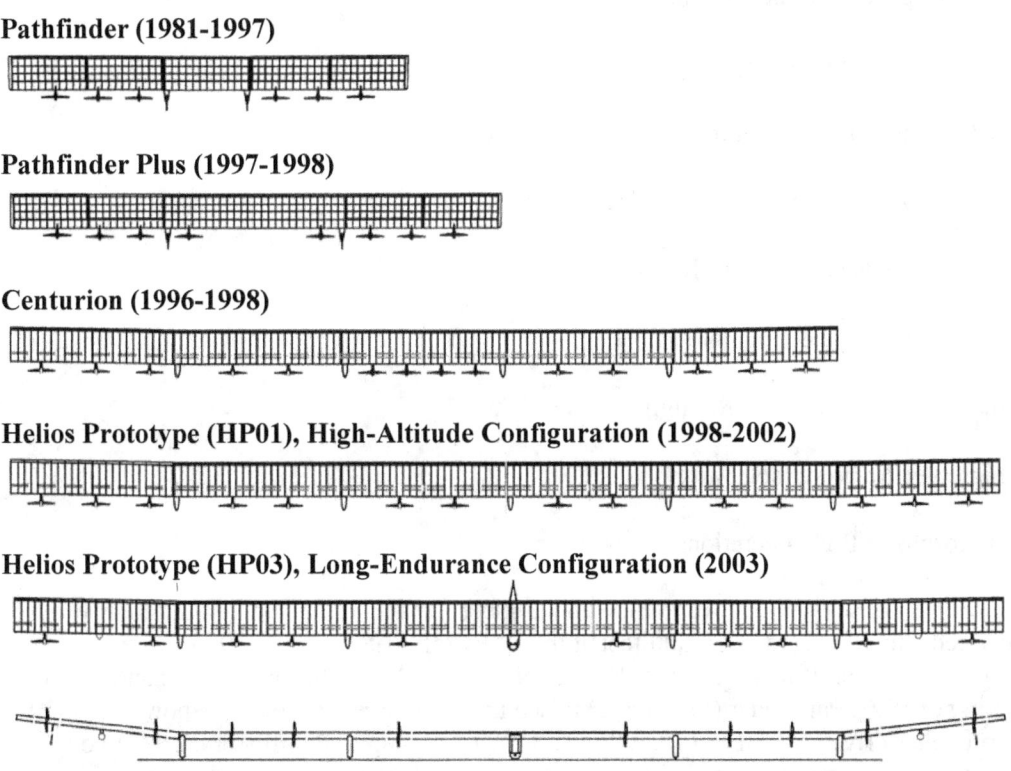

Figure 6.1 - Solar Aircraft Evolution through the ERAST Program

In the next few paragraphs, each vehicle listed in Figure 6.1 and the significant accomplishments associated with those vehicles are briefly reviewed to provide a clear understanding of how the ERAST program and the HALE vehicles evolved from 1994 into 2003, and what were the circumstances that ultimately led to a decision to fly the HP03 aircraft on 26 June 2003.

Pathfinder: The first generation HALE vehicle was the Pathfinder, a flying wing with a wingspan of about 100 feet powered by six battery-operated electric motors. The vehicle had two underwing pods, which contained the landing gear, the batteries, the instrumentation system, and the flight control computer. Pathfinder was the technology test bed for developing many of the enabling technologies and processes for solar-powered stratospheric flight. These technologies included:

- lightweight composite structures

- low wing loading flying wing

- redundant and fault tolerant flight control system

- lightweight and low power avionics systems

- low Reynolds number aerodynamics

- high efficiency electric motors

- thermal control systems for high altitude flight

- high specific power solar array

- stratospheric flight operations

With the addition of solar cells covering the entire upper surface of the wing, Pathfinder set a solar-powered altitude record of 50,500 feet at the DFRC on 11 September 1995. After further modifications, the aircraft was moved to the U.S. Navy's PMRF on the Hawaiian island of Kauai. In the spring of 1997, Pathfinder (Figure 6.2) raised the altitude record for solar-powered aircraft, propeller-driven aircraft, to 70,500 feet. During this flight, Pathfinder carried two lightweight imaging instruments to learn more about the island's terrestrial and coastal ecosystems, demonstrating the potential of such aircraft as platforms for scientific research.

Figure 6.2 - Pathfinder During its Altitude Record Setting Flight

Pathfinder Plus: The Pathfinder Plus vehicle was the next step leading to the Helios Prototype. The Pathfinder aircraft was enlarged to a 120-foot wing span aircraft by using four of the five sections from the original Pathfinder wing and a new 40-foot center wing panel section. This center wing section was of the same design as the wing section of Pathfinder Plus' successor, the Centurion, which was being designed to reach 100,000-foot altitude on solar power. In addition, the number of electric motors was increased to eight. The Pathfinder Plus allowed flights to higher altitudes and was used to flight qualify the Centurion wing panel structural design, airfoil, and SunPower solar array. Three Pathfinder Plus flights were conducted at the PMRF. The final flight on 6 August 1998 achieved a new record altitude 80,200 feet. These flights validated the power, aerodynamic, and systems technologies needed for the Centurion.

Centurion: Development of the Centurion aircraft, the third generation, began in late 1996. Originally, the ERAST goals were to build two airframes: one for demonstrating a Centurion high-altitude (100,000-foot altitude) mission and one for demonstrating a Helios long-endurance (96 hours at 50,000-foot altitude) mission. To begin addressing the first goal, a 1/4-scale version of the Centurion aircraft was designed, built, and flight-tested to verify a new high-altitude aerodynamic airfoil design and to evaluate aircraft handling qualities. Also, all of the key technologies that were developed on Pathfinder were further improved into lightweight, more efficient, and more robust subsystems. In 1998, the full-scale Centurion (Figure 6.3) was built. The vehicle had five wing panels with a total wingspan of 206 feet, 14 electric motors to provide level flight at 100,000-foot altitude, and 4 underwing pods to carry batteries, flight control system components, ballast, and the landing gear. In late 1998 the Centurion flew three development test flights at the DFRC at low altitudes using battery power to verify the design's handling qualities, performance, and structural integrity.

Figure 6.3 - Centurion During its Low Altitude Flights

Helios Prototype HP01 (High-Altitude Configuration): In early 1999 under the constraint of a reduced budget that could fund only one aircraft, NASA and AV agreed the best way to proceed was to use a single airframe to demonstrate both of the ERAST goals. Based on this plan and to demonstrate the ERAST goal of sustained flight near 100,000 feet, the Centurion was modified from a 5-wing panel to a 6-wing panel aircraft by replacing the center wing panel with two new stronger center wing panels and by adding a fifth landing gear. This change resulted in the wingspan being increased to 247 feet. The aircraft continued to use 14 electric motors, with the four center motors redistributed on the new center wing panel. Following these modifications the name of the aircraft was changed from Centurion to the Helios Prototype, thus becoming the fourth configuration in the series of solar-powered flying wing demonstrators.

Using a traditional incremental approach to flight testing, the Helios Prototype (HP99) was first flown in a series of six battery-powered, low altitude, development flights in late 1999 at DFRC to validate the longer wing's performance and the aircraft's handling qualities. Various types of instrumentation required for the planned solar-powered high altitude and long endurance flights

20

were also checked out and calibrated during these initial low-altitude flights. Four flights were conducted to assess the high-altitude configuration and two flights, with the aircraft ballasted for the "then" planned regenerative fuel cell system (RFCS) hardware and the solar array, were conducted to assess the performance of the heavier configuration. At this time the long endurance configuration was intended to use only eight electric motors.

In 2000-2001, the HP01 was upgraded with new avionics, high altitude environmental control systems, and a new SunPower solar array (62,000 solar cells), and on 13 August 2001 flying out of the PMRF, the aircraft (Figure 6.4) reached an altitude of 96,863 feet, a world record for sustained horizontal flight by a winged aircraft.

Figure 6.4 - HP01 High-Altitude Configuration

Fuel Cell Development: In late 1998, NASA and AV started the preliminary design and development of the RFCS for the long endurance demonstration planned for 2003. A conceptual design review for the aircraft with a RFCS was held in May 1999 and a preliminary design review (PDR) was held in September 1999. NASA and AV committed to the development of the RFCS in October 1999 and, soon after, AV and two fuel cell subcontractors started the development of the RFCS. By the summer of 2001, a prototype, full-scale fuel cell pod was built, but the hydrogen-oxygen fuel cells and electrolyzers under development were not working reliably. It was clear at that time that designing, building, and testing two flight weight RFCS pods for the long endurance demonstration would not be possible with the time and budget remaining to the program.

21

During October-November 2001, NASA and AV sponsored independent technical reviews to assess the progress in the RFCS development. Based on these assessments, AV approached DFRC management with a proposal to change from a RFCS to a consumable primary fuel cell system (PFCS). The primary motivation for the proposed change was two-fold: 1) a PFCS, derived from existing fuel cell components in the automotive industry, could be designed, built, and tested within the current schedule and budget constraints, and 2) a Helios UAV with a PFCS would have a 7-14 day duration capability. This latter factor was important because AV thought that they could attract other commercial and Department of Defense (DOD) customers and bring this HALE capability to market sooner. In December 2001, NASA and AV decided to switch to the PFCS and began the development and modifications needed to the HP01 aircraft for the 2003 demonstration. Since 2003 was the last year of the ERAST program, a major milestone had to be accomplished without the possibility of schedule or budget relief. This contributed to the decision to switch to the primary fuel cell as a risk reduction. It also made it harder to consider other risk reduction efforts such as a low-altitude flight test at the DFRC.

Helios Prototype HP03 (Long-Endurance Configuration): The primary objective of the 2003 flight test program was to use a hydrogen-air fuel cell to sustain flight overnight at 50,000 feet. The aircraft to be used for the long endurance demonstration in 2003 was designated the HP03. The PDR and Critical Design Reviews (CDR) for this aircraft were held in February 2002 and in August 2002, respectively. In addition, separate independent technical reviews were conducted from November 2002 through January 2003 to review the aircraft configuration changes, structural loads, stability and control, and aeroelastic models and predictions. Based on these design reviews, a decision was made to strengthen the wing tip spars so that their structural margins would be consistent with the structural margins along the rest of the wing spar under the design load conditions.

NASA and AV also recognized that the structural, stability and control, and aeroelastic margins of safety were less on the HP03 than on the HP01. However, these margins were still sufficient to conduct the 2003 long endurance flight demonstration. It was also recognized that the mass distribution for HP03 was significantly different than the mass distribution of the initially proposed demonstrator with a RFCS system. The aircraft with the RFCS would have required only two regenerative fuel cell pods located at about 1/3 the distance from the vehicle centerline to the wing tip. The aircraft with the PFCS installed was more point loaded in that 3 pods were required. The heavy primary hydrogen-air fuel cell pod (520 lbs) was located at the centerline of the aircraft and the 2 high-pressure hydrogen fuel tanks (165 lbs each) were located at the center of each wing tip panel. A schematic of the PFCS is provided in Figure 6.5.

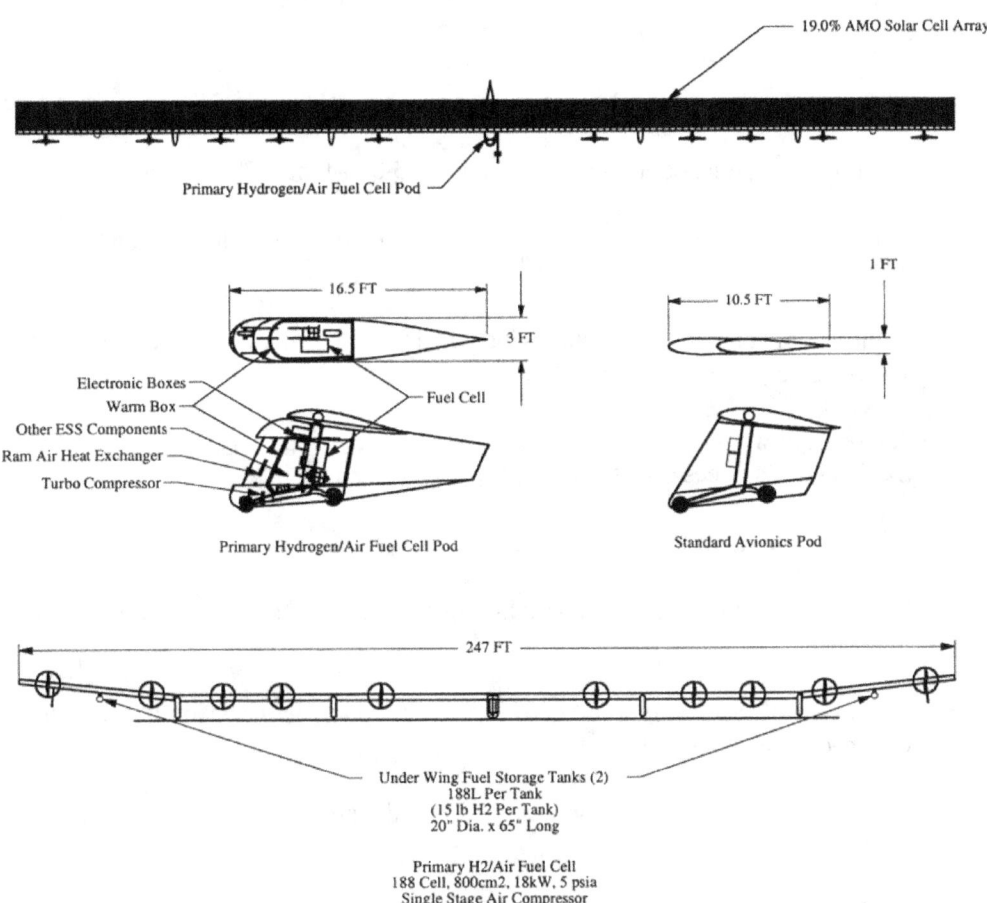

Figure 6.5 - Schematic of HP03 Hydrogen-Air Fuel Cell Configuration

Aircraft Modifications Following HP01 Flights: By the end of 2002, the PFCS was designed and fabricated, and the aircraft was modified. Primary modifications to the aircraft included:

- Center pod was replaced with a fuel cell pod weighing approximately 520 lbs;

- Two hydrogen fuel tanks weighing approximately 165 lbs each (including 15 lbs of hydrogen) were added beneath the wings on the wing tip panels at motor pylon locations #2 and #13. Hydrogen supply lines and data lines were added between these tanks and the fuel cell pod;

- Motors on pylons #2, 6, 9, and 13 were removed resulting in 10 motors;

23

- Spar strengthener in the form of a concentric tubular inner spar was added to the tip panel spars;

- Aluminum center joiner tube was replaced with a lighter weight carbon fiber tube;

- Pathfinder propellers optimized for flight at 65,000-foot altitude were installed;

- Tip panel incidence was reduced to 0 degrees; the high altitude configuration had 1 degree of incidence on the tip panels;

- Front row of solar cells on center wing panels and the first two front rows from mid and wing tip panels were removed;

- Servos from wing tip panels were removed and the wing tip panel elevators were fixed at –2.5 degrees offset (trailing edge up);

- Wing tip landing gear was installed;

- Hydrogen feed lines from outboard hydrogen tanks to center fuel cell pod were installed;

- Flight control system autopilot gains were revised and gain scheduling with airspeed was made available;

- Three battery packs were reconfigured into pod 2 and pod 4 to mass balance the aircraft.

By April 2003, testing of the PFCS was completed and integrated into the aircraft, and all combined systems tests were accomplished. The final gross weight for the HP03 was 2,320 lbs as compared to the 1,585 lbs gross weight of HP01 during its altitude record flight in 2001, an increase of 735 lbs.

HP03 Characteristics: The HP03 load carrying structure was constructed mostly of composite materials. The main wing spar was made of carbon fiber, was thicker on the top and bottom to absorb the bending that would occur during flight, and was wrapped with Nomex and Kevlar to provide additional strength. The wing ribs were made of epoxy and carbon fiber. The wing leading edge consisted of aerodynamically shaped styrofoam, and the entire wing was wrapped with a thin, transparent plastic skin. As described earlier, the aircraft consisted of 6 panels for a total wingspan of about 247 feet. Aerodynamically shaped underwing pods were attached at each wing panel joint to carry the landing gear, the battery power system, the flight control computers, and flight instrumentation. The wing had no taper or sweep, an 8-foot wing chord (aspect ratio of 31) with a maximum thickness of 11.5 inches (constant from wingtip to wingtip), and 72 trailing-edge elevators spanning the entire wing.

The aircraft was powered by 10 brushless direct-current electric motors rated at 2 hp or 1.5 kW each. The two-bladed propellers were 79 inches in diameter, made of composite materials, and designed for high efficiency at high altitudes. To turn the aircraft in flight, differential power was applied to 8 of the 10 motors (power to the outboard 4 motors on one wing was increased while power to the 4 motors on the other wing was decreased). Servomotors commanded by the aircraft's flight control computer drove the trailing edge elevators for pitch control. To provide adequate lateral stability, the outer wing panels had a built-in 10-degree dihedral (upsweep), and to prevent wingtip stall during the slow landings and turns, the wing tip had a slight upward twist.

Straight Line Flight – 15 May 2003: On 15 May 2003 the project team conducted a successful straight-line flight test and mission dress rehearsal in preparation for the first high-altitude flight in June. The primary objectives of the straight-line flight were to verify the proper wing dihedral distribution of Helios by flying a short hop with a straight-ahead landing, and to conduct all of the necessary preflight assembly and test procedures required for a high-altitude mission. The aircraft was flown at an altitude of 2 feet above the runway for about 10 seconds. The assessment of the flight results indicated the aircraft had approximately the correct wing dihedral distribution, and that all of the aircraft systems, the fuel cell pod, and the ground support equipment were working well with the exception of the solar array. Test data from this flight allowed for fine-tuning the aircraft's mass distribution, wing tip panel incidence, elevator settings, and flight control system gains to help establish a safe operating envelope for high altitude flight investigations. Prior to the first high altitude flight scheduled for June, several minor modifications were made to the aircraft. These modifications included:

- New main battery pack installed

- Failed motor replaced

- Fuel cell pod tail fairing repaired

- Broken bus bars on the solar array repaired

- Flow meter and two pressure sensors in the fuel cell pod replaced

- Autopilot gains revised

First Flight (HP03-1) – 7 June 2003: On 7 June 2003 the first flight of the aircraft, designated HP03-1 (Figure 6.6), was accomplished. The objectives of this flight were to:

- Demonstrate the readiness of the aircraft systems, fuel cell systems, flight control system, flight support equipment, range support instrumentation, and procedures required for a long duration flight;

- Validate the handling and aeroelastic stability of the aircraft with its fuel cell system and gaseous hydrogen storage tanks installed;

25

- Demonstrate the operation and the performance of the fuel cell system in the stratosphere;

- Provide flight, fuel cell pod, and ground crew qualification training for the additional personnel required to staff future multi-day flights.

The flight performance predictions estimated that the HP03 was capable of approximately 30-hour flight duration at 50,000 ft altitude. During the flight, data were measured in real-time to validate the predicted aeroelastic characteristics of the aircraft and to demonstrate that the vehicle was aeroelastically stable at the flight conditions expected for the long-endurance flight demonstration. The aircraft flew flawlessly thus validating the handling and aeroelastic stability of the aircraft in smooth air. However, the flight was aborted about 15 hours after takeoff because of some leakage associated with the coolant system and compressed air lines that feed the PFCS. Because of this leakage, the test team was unable to start-up the fuel cell system.

Figure 6.6 - HP03-1 Flight on 7 June 2003

Aircraft Modifications Following First Flight: The turbulence levels and winds during the first flight were uncharacteristically light. As a result there was concern that the airspeed variations in turns, the high sideslip at low-power/low-altitude conditions (i.e. landing), and the sensitivity of wing dihedral to power setting over the entire flight envelope would make the aircraft more difficult to handle and to land safely under the more normal weather conditions that would be present in the flight test area. To address these concerns the project team elected to make changes to the HP03-1 aircraft prior to the next scheduled flight (26 June 2003).

Major changes made to the aircraft to improve the aircraft handling qualities, to reduce wing dihedral sensitivity to power setting, and to increase wing dihedral at low power included:

- Motor propeller pitch was flattened from –5.5 to –8 degrees;

- Power throttle scaling on the two outboard motors was reduced to 50 percent of the center motors;

- Drag mode on the tip motors was eliminated;

- Wing tip panel incidence angle was increased from 0 to 0.5 degrees.

- Flight control system autopilot longitudinal gains were increased by 3db;

- Ratio of the airspeed hold gain to the pitch attitude damping gain was increased by a factor of 2;

- Longitudinal gain switch on the pilot's controller multiplier was reduced;

- Limiter on the value of the airspeed error integral was increased.

Second Flight (HP03-2) – 26 June 2003: The second flight of the aircraft, designated HP03-2, took place on 26 June 2003. The objectives of the test were to:

- Clear the aircraft flight envelope for the new aircraft configuration changes, and for the 50,000 feet to 60,000 feet altitude climb/glide needed for the planned long-duration mission;

- Verify stable operation of the fuel cell and compressor at an altitude of 50,000 feet;

- Achieve fuel cell pod rated flight power of 18.5 kW at 50,000 feet;

- Run the fuel cell pod system for at least 2 hours to develop confidence that it can run all night;

- Perform a modest fuel cell performance sensitivity matrix so that the results can be used to optimize the performance for a long-duration mission;

- Demonstrate a rapid shutdown of the fuel cell pod and night restart on battery power.

A precise description of the second flight, the flight involving the mishap, is provided in Section 7.

Some Summary Comments

Table 6.1 provides a summary of key events beginning with the decision to demonstrate both a high-altitude mission and the long-endurance mission using one vehicle.

Table 6.1 - Chronological Order of Key Events in Preparing for the HP03

- Late 1998: Initiated design of a RFCS for the planned 2003 long endurance HP03 flight demonstration;

- Early 1999: NASA and AV agreed to use a single airframe to demonstrate the ERAST goals of high altitude and long endurance flight;

- May 1999: Completed CDR for the aircraft with a RFCS;

- September 1999: Completed PDR for the aircraft with a RFCS;

- October 1999: NASA and AV committed to the development of the RFCS;

- Late 1999: HP01 first flown in a series of six battery-powered, low altitude, development flights at DFRC;

- 2000-2001: HP01 upgraded with new avionics, high altitude environmental control systems, and a new SunPower solar array;

- Summer 2001: Prototype RFCS built;

- August 2001: HP01 reached an altitude of 96,863 feet, a world record for sustained horizontal flight by a winged aircraft;

- October-November 2001: NASA and AV sponsored independent technical reviews to assess the progress in the RFCS development;

- December 2001: NASA and AV decided to switch to a PFCS concept and initiated the development and modifications to the HP01 aircraft;

- February 2002: Completed PDR for HP03 completed;

- August 2002: Completed CDR for HP03 completed;

- November 2002-January 2003: Independent technical reviews completed to review HP03 changes, including structural loads, stability and control, and aeroelastic models and predictions;

28

- January 2003: Design, development, and fabrication of the primary fuel cell pod, and the HP03 aircraft modifications completed;

- February 2003: PFCS testing and aircraft subsystem functional testing completed;

- 12 February 2003: Mission Success Review (MSR) held with NASA and AV to review the final aircraft configuration changes, flight operations approach, risk management process, systems safety, and program controls;

- March 2003: PFCS pod integrated to the fully assembled aircraft in Simi Valley;

- 12 March 2003: First Deployment Readiness Review (DRR) was conducted to assess the system design modifications, qualification testing, airworthiness, and operational readiness for the hydrogen tanks and ground support equipment;

- 13 March 2003: Aircraft combined systems testing completed to verify the transition from PFCS pod power to main battery power;

- 2 April 2003: Aircraft shipped to the PMRF;

- 8 April 2003: Second DRR conducted to assess the system design modifications, qualification testing, airworthiness, and operational readiness of the aircraft and the PFCS pod;

- 14 April 2003: Aircraft arrives at the PMRF;

- 15 - 22 April 2003: Hydrogen tanks and support equipment shipped to the PMRF;

- April 2003: PFCS pod shipped to the PMRF;

- 27 April 2003: PFCS pod arrives at the PMRF;

- 28 April 2003: PFSC pod inspected, integrated to the aircraft, and functionally tested;

- 1 May 2003: Remote hydrogen filling station set-up and activated;

- 3 May 2003: Completed 9-hour aircraft systems test;

- 15 May 2003: Straight-line flight test completed to verify the proper wing dihedral distribution of aircraft, and to assess all of the necessary preflight assembly and test procedures;

- 30 May 2003: Technical Briefing with NASA and AV personnel conducted to review the ground and straight-line flight test results, flight plan, configuration changes, operational procedures, and system safety, and to close out any remaining items prior to flight;

- 5 June 2003: Helios H03-1 crew briefing held; reviewed test objectives, final aircraft status, test timeline, safety issues, roles and responsibilities, flight plan, and weather forecast;

- 7 June 2003: First high altitude flight of the HP03 conducted to validate the handling and aeroelastic stability of the aircraft with its fuel cell system and gaseous hydrogen storage tanks installed, and to obtain data on the performance of the fuel cell system in the stratosphere;

- 24 June 2003: Technical briefing with NASA and AV personnel conducted to review the HP03-1 test results, flight plan, configuration changes, operational procedures, and system safety, and to close out of any remaining items prior to flight;

- 24 June 2003: Crew briefing (see Appendix F, Document F.6) held with a Go/No-Go poll; all NASA, AV, and PMRF personnel supporting the flight test indicated a "Go" for flight on June 26;

- 26 June 2003: Second high altitude flight of the HP03 conducted.

Section 7
Narrative Description of Mishap

This section of the report presents a very precise accounting of the project team activities beginning late in the evening of 25 June 2003 and leading up to the mishap at 10:36am on 26 June 2003. Document C.1 (Appendix C, Volume II) provides photographs of the vehicle and supporting equipment from pre-flight preparation through recovery of the wreckage.

Sequence of Events Up to Mishap
Preflight: At 11:30pm on 25 June, the 10-1/2 hour Helios mission countdown started with towing the aircraft halves out of the hangar, mating the two halves, installing both hydrogen tanks on the outboard wing tip panels, powering up the aircraft and stationary Ground Control Station (GCS), conducting radio frequency (RF) data link tests with the stationary GCS, and conducting a fuel cell pod preflight test.

At 4:56am, the countdown transitioned to powering up the mobile GCS, conducting mobile GCS data link tests, autopilot loss of link (LOL) tests, and switching aircraft power from ground power to auxiliary battery power. All aircraft and fuel cell systems were performing well during pre-flight testing. The local weather prediction for sunrise was light winds out of the east and westward moving patches of low clouds. Upper level cirrus clouds had moved over the Hawaiian Islands overnight, but were continuing northward, and forecasted to be out of the area by take off. Figure 7.1 shows HP03-2 positioned on the transient ramp at sunrise waiting to be towed to the solar runway.

Figure 7.1 – HP03-2 Awaiting Takeoff

31

At 6:17am, the aircraft was towed to the solar runway. While being towed, a Go/No-Go review for the flight was held in the stationary GCS with the AV flight crews and NASA. Based on the results of all of the preflight tests, compliance with mission rules, and the weather forecast, the team was a "Go" for flight. However, during the poll the weather forecaster indicated a "*very marginal GO*". Even though all the written weather constraints would be met by takeoff time, he was concerned that the close proximity of the southern shear line would create turbulence during climb out and with the jet stream turbulence near 35, 000 feet as suggested by the billows in the cirrus clouds.

At 6:48am, the final aircraft preflight checks were started. All checks were nominal.

At 7:50am, the aircraft was unloaded from the transport dolly and positioned for a southbound takeoff based on the forecast winds for a takeoff window of 8:30am to 10:30am. The initial motor power run-up test was marginal due to the upper-level cirrus clouds blocking the sunlight. Subsequently, the clouds moved to the north and a repeated test was successful.

At 8:52am, the new weather forecast indicated that the airfield winds would not switch to a southerly component as predicted but would continue coming from the northwest. After some discussion with the mobile and stationary flights crews, the Mobile Flight Director (MFD) decided to reposition the aircraft for a northbound takeoff. The aircraft was placed back on the dolly and moved to the south end of the runway.

By 9:30am, a quantity (sigma w) measured directly by the Sodar that is proportional to the strength of small-scale vertical velocity fluctuations and indicative of low-level turbulence risk, had increased from 0.4 m/sec to as high as 0.8 m/sec. Although these sigma w values did not violate mission rules of 1.0 m/sec, the Helios Meteorologist did advise the MFD that these values were the largest he'd seen for a takeoff, and high enough to expect light to moderate turbulence. After 9:38am, the sigma w value edged down to 0.7 m/s. In addition at 9:36am, the tower reported seeing white caps building offshore, indicative of the proximity of the shear lines. Whitecaps are clearly visible in Figure 7.2.

At 9:45am, the aircraft was unloaded from the dolly and the final preflight tests were completed. The takeoff was delayed by approximately 20 minutes while the Mobile Pilot (MP) waited for clouds near the airfield to open up providing full solar illumination on the HP03-2 solar array and a clear path for a westerly climb-out from the airfield. Low and high altitude weather conditions were within the weather element limits and were characterized as typical weather conditions for a solar aircraft flight at PMRF with the exception of the high sigma w values and the proximity of the southern shear line. A valid flight plan showed sufficient operating margins for the entire flight.

Takeoff: At 10:06am, about 1 1/2 hours later than planned, with winds at 7 knots from 300 degrees and scattered cumulus clouds, the MP throttled up to 50% (16 kW) for takeoff. Clouds were shadowing parts of the runway as the aircraft lifted off (Figure 7.2). The airplane headed out to sea as usual, however the climb rate was slightly less than normal due to cloud shadows.

For the first ten minutes the stationary crew worked primarily at helping the mobile crew navigate around the clouds.

Figure 7.2 - HP03-2 on Takeoff

At 10:19am, the Stationary Pilot (SP) noted that there was indication that the airplane was in turbulence.

At 10:22am and again at 10:24am, the aircraft's wing dihedral became larger than normal. In both cases a mild pitch oscillation occurred. The wing dihedral returned to normal, and the oscillation damped out. These events both occurred during the six minutes when both crews were focused on the handoff procedure. Neither crew was aware of the high wing dihedral or the pitch oscillations.

At 10:23am, the N716AM (helicopter chase aircraft) pilot relayed a suggested 40° right turn to move away from whitecaps in the ocean that appeared to be associated with the island's southern shear line and to find smoother air. The MP concurred and turned.

Transfer of Control from MP to SP: At 10:25am, handoff of aircraft control from the MP to the SP was completed. The actual handoff occurred during the second of the pitch oscillations described earlier. Following the hand-off, the aircraft was at 37 ft/sec commanded airspeed. The throttle command was set to maximize the solar array power without dipping into the batteries at approximately 122 volts. The climb rate of the aircraft varied between about 85 and 100 ft/min. Figure 7.3 shows the aircraft as it continued to climb.

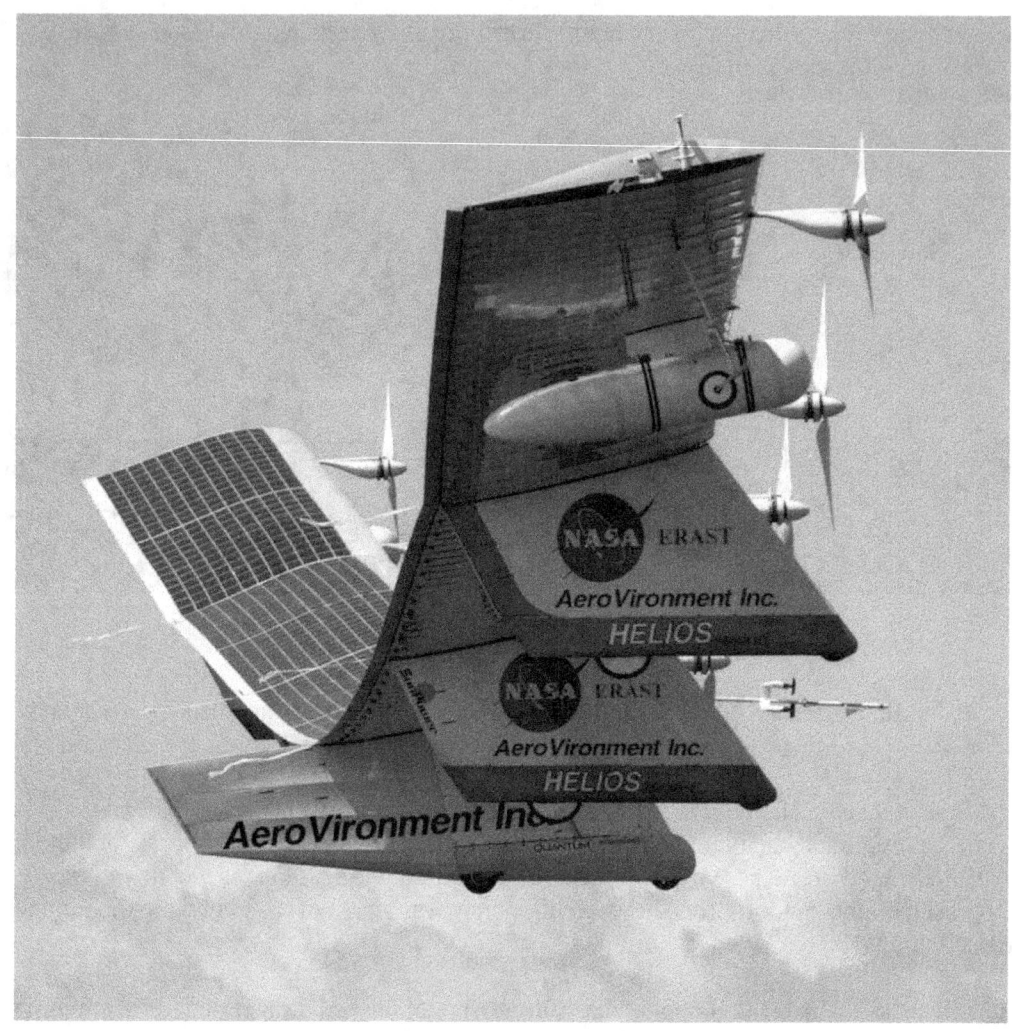

Figure 7.3 - HP03-2 Close-up During Climb

At 10:28am, the helicopter observer advised a 30-degree left turn to avoid the northern shear line. The SP announced a 15-degree left turn to stay closer to the planned flight path. The actual turn was about 20 degrees.

At 10:30am, the Planning Engineer (PE) advised the SP to maintain a northerly component in the flight path to avoid getting too close to the southern shear line. Agreement was reached to continue the present heading.

At 10:34am, the helicopter observer advised again that the aircraft was approaching the northern whitecaps in the ocean, an indication a wind shear line was being approached. The helicopter observer suggested a left turn of 20 degrees. The SP responded with a 17-degree left turn.

The Mishap

The SP had selected the wing-tip video camera to observe the aircraft because he believed it would give the best indication of wing dihedral and also how the aircraft was responding to any turbulence. At 10:35am, at approximately 2,800 feet altitude as defined by the Global Positioning System (GPS), the aircraft began experiencing airspeed excursions of around ±2 ft/sec. At this time the wing-tip video indicated that the dihedral seemed high for the speed the aircraft was flying. When looking at the video, the camera view essentially pointed at the top of the center of the aircraft rather than the more usual view looking across the wingspan. The high dihedral persisted (Figure 7.4), and a pitch oscillation built up, similar to what had previously occurred but not observed and thus had not been interpreted as a potentially dangerous periodic oscillation. The SP noted that he was continuing to see large wing dihedrals and asked the Stability and Control Engineer (S&CE) to confirm that the correct procedure to reduce dihedral was to increase airspeed. The S&CE concurred, and at 10:35:12am the SP selected an increase in airspeed of 1ft/sec (37 ft/sec to 38 ft/sec). The dihedral decreased slightly then grew past 30 feet.

Figure 7.4 - HP03-2 at High Wing Dihedral

At 10:36:03am, the SP noted large airspeed fluctuations, an indication that the aircraft was experiencing large pitching motions, and asked the DE for suggestions to stabilize the aircraft. The airspeed excursions were about ±10 ft/sec and diverging. At this point the wing-tip video began to show large pitch excursions as the horizon was coming in and out of view. The SP noted at 10:36:17am that he thought the airplane was in a large phugoid oscillation in that the airspeed excursions were almost off the scale. In fact the amplitude of the unstable pitching mode about doubled every cycle. At 10:36:23am the SP initiated the non-deferred emergency

35

procedure for a pitch hard over and immediately turned the Airspeed Hold off. As Airspeed Hold was being turned off, the vehicle was already pitching down sharply and accelerating to an airspeed approximately two-and-a-half times the maximum design airspeed.

At 10:36:28am at these extreme conditions, the aerodynamic loads broke the leading edge foam sections of the right wing near the hydrogen fuel tanks and then began ripping the solar cells and skin off the upper surface of the wing. The wing-tip video continued to function for a few more seconds and showed other portions of the aircraft starting to tear away as the aircraft continued to destruct. The last video frames show the wing sweeping aft. This could be due to the fuel cell weight at the aircraft centerline and increased drag outboard due to loss of the skin. Figure 7.5 shows a photograph of the aircraft as it is falling towards the ocean.

Figure 7.5 – HP03-2 Falling Towards the Pacific Ocean

At 10:37am, the aircraft impacted the ocean in mile-deep water 10 miles off the coast of Kauai, where, upon impact, the structure including the main load carrying composite wing spars were severely damaged (Figure 7.6). The elapsed time from the first effort to diagnose and correct the high wing dihedral condition to the point at which the airplane began to break up in the air was 91 seconds. Figure 7.7 provides time histories of the aircraft's pitch rate and airspeed, and the wing dihedral for the 30-minute flight.

Figure 7.6 – HP03-2 upon Impact with the Ocean

Helios Flight H03-2

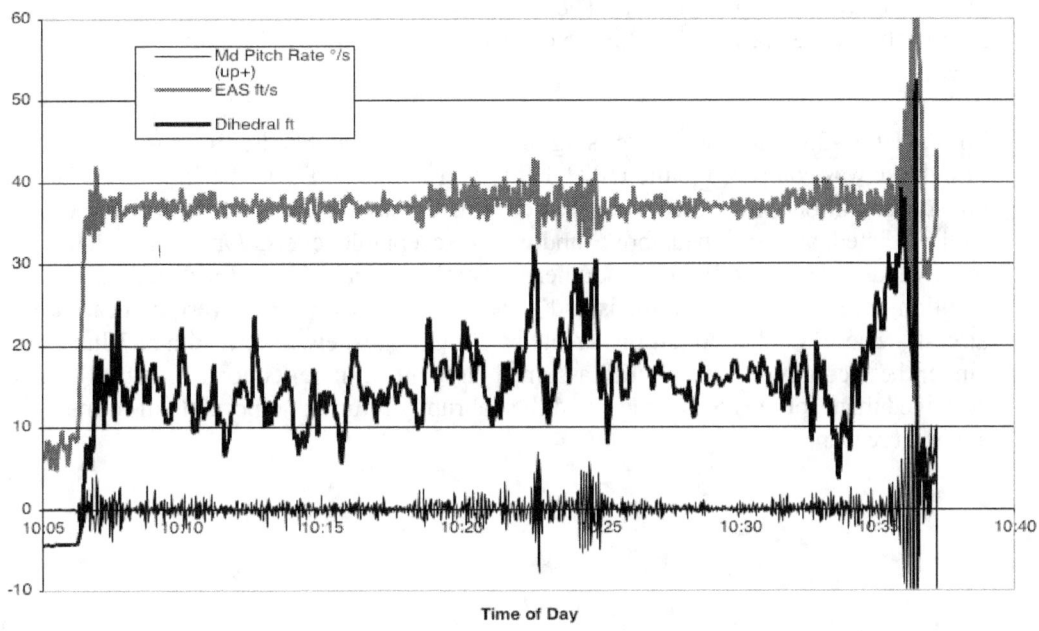

Figure 7.7 – Wing Dihedral, Pitch Rate, and Airspeed History for Flight HP03-2

Other than damage to the aircraft, there was no property damage on the ground or injuries to personnel as a result of the mishap. Salvage teams consisting AV and NASA personnel recovered most of the largest pieces (75 percent of the aircraft by weight) of the vehicle with the exception of the heavier fuel cell system, which sank into the ocean. The recovered wreckage was shipped to California and, later, parts of the right main wing spar were transported to LaRC for further study.

Some Summary Comments

Analyses performed prior to the mishap accurately predicted the wing dihedral shape in smooth air. These analyses also predicted that the aircraft would be unstable for a wing dihedral greater than about 30 feet. However, these analyses did not predict the degree of aircraft's increased sensitivity to disturbances like turbulence, the inability of the aircraft to restore itself to some nominal dihedral after being disturbed, or the highly divergent nature of the unstable pitch mode. Actual events related to these predictions are described below:

- HP03-2 encountered significant turbulence, presumably near the island shear lines on four occasions (see Figure 2, Appendix D.1.e) during the 30-minute flight. The first event did not result in the development of high dihedral. The next two, following shortly after the first event, occurred in the space of three minutes. The aircraft developed high dihedral deflections of about 30 feet, began to oscillate in pitch (as the analyses had predicted), and recovered on its own without pilot interaction. The flight crews (Mobile and Stationary) did not interpret the airspeed variations associated with the events as periodic oscillations, but rather as typical responses due to turbulence. On the third occasion, the dihedral increased to an even higher magnitude of about 40 feet, and the pitch oscillation diverged rapidly. The loss of pitch control led to an over-speed of the aircraft that caused progressive failure of secondary structure and finally the loss of the aircraft.

- For the HP03-2 flight, it was surprising just how strong the dihedral response to turbulence was compared to the HP03-1 flight on 7 June 2003. In the smooth air (an unusual condition for the area) observed on 7 June 2003 the aircraft dihedral shape was well predicted, as mentioned above, and within acceptable levels. On 26 June 2003 the trade winds veered slightly more southerly than the typical east-northeast direction. In addition, the shear lines from the island's wake boundaries were converging near the aircraft. These weather conditions created a stronger gust environment than 7 June. Under these conditions, the aircraft assumed large dihedral deflections (first two events), sustained them for tens of seconds, and then abruptly returned to normal, all for no apparent reason.

- It was also observed that the aircraft dihedral shape in turbulence was not robust, that is, the aircraft was slow to restore to a more classical and stable shape. Although analyses performed before the mishap flight showed this fact, the first flight on 7 June appeared to confirm that the less robust configuration could be made acceptable with appropriate control system gain changes. This may have been false confidence in that flight HP03-1 was flown in one of the most benign turbulence conditions encountered during solar-aircraft flights from PMRF.

- As mentioned above, the rapid divergence of the pitch oscillation when the dihedral reached 40 feet (the third event) was not expected. Analysis had predicted that the aircraft was not stable in pitch for dihedral greater than 30 feet, but it was not anticipated that the divergence would be so rapid. Pitch oscillations encountered in previous flight tests of vehicles of the Helios class were mild, giving the crew time to deliberate on a course of corrective action. For all of the past events, the pitch oscillations quickly damped out when the corrective action was taken. The predicted pitch instability was considered acceptable, because there was no history (or prediction) of large, sustained dihedral deflections.

- In earlier years the strategy on climbout was to avoid the shear lines until the airplane achieved sufficient altitude to get above the most significant turbulence. Aircraft measurements and flow simulations in the area of the mishap indicate that this altitude is 4000 to 5000 feet. Because HP03 flew at a somewhat higher equivalent airspeed than previous solar-powered configurations without an increased rate of climb, the slope of the climbout trajectory was smaller than previous flights. This meant a longer flight trajectory over which the airplane was exposed to the greater turbulence at lower levels, and, at the same time, trying to avoid the shear lines.

Section 8
Independent Analyses and Tests

Independent Working Groups performed several studies to support the MIB. The studies and results are described in documents provided in Appendix D, Volumes II (nonproprietary reports) and III (proprietary reports). This section of the report provides a brief summary of the IWG investigations.

Structural and Control System Modeling Working Group

Appendix D.1.a (Volume III)
Investigation of Nonlinear Structural Contributions to the Helios Mishap
The purpose of this study was to identify potential nonlinear structural contributions to the Helios mishap. These analyses assumed that the structure was made of linear elastic materials, but having a nonlinear strain-displacement relationship. Thus, only the geometric nonlinearities were considered in the analyses; material nonlinearities associated with material failure were not included.

There were three separate nonlinear analyses performed. The first analysis involved converting the linear finite element (FE) analysis input files created by AV to nonlinear analysis FE input files. These calculations demonstrated that both linear and nonlinear analyses with a uniform running load could predict high wing tip deflections (dihedral) in the range of interest.

The second analysis involved building a representative wing panel section using proprietary data from AV to identify potential bending-torsion coupling effects. Based on the available data, uniform properties were generated for the leading and trailing edge spars, and rigid properties were assumed for the central and end ribs. Based on the properties and the panel section modeled, there was no identified bending torsion-coupling, however, the analyses predicted that for extreme deflections, the cables would overcome the pre-tensioning and go 'slack'. When the cables slacken, a difference in the leading and trailing edge spar deflections was observed. Both the moment levels and deflections were predicted to be significantly beyond the design levels of the structure.

The third analysis was performed to identify whether the nonlinear response of the spar shell under moment would deviate from the linear response and soften. Using a shell model with composite properties the nonlinear analysis found no ovalization or nonlinear effects outside of the load introduction region.

The study concluded that for the limited analyses performed no conclusive nonlinear structural effects could be identified that contributed to the Helios mishap.

Appendix D.1.b (Volume III)
Aeroservoelastic Stability Analysis of Helios Fuel Cell Aircraft
This report describes an aircraft trim modeling activity that attempted to combine a structural model developed using NASTRAN with an aerodynamics model developed using ZAERO. The

primary purpose of this effort was to corroborate the ASWing analyses performed at AV, and to identify elastic deformations that might explain the relatively sluggish response of the wing to reduce wing tip deflection once disturbed to relatively high dihedrals.

The study was not completely successful in modeling the observed Helios trim behavior. Although the model produced a "gull winged" Helios trim shape similar to that predicted by ASWing, the shape was never observed in flight. Additionally, although the NASTRAN ZAREO predicted a trim solution to a high dihedral configuration after just four iterations, the solution would then begin to diverge. This behavior in the numerical behavior was interpreted as a non-unique solution. Nonetheless, the stability analysis performed for the configuration showed an over-damped response to external loading. In this state the structure would resist deflecting back to some nominal dihedral. The wing tips would be sluggish in reducing the deflection angle once perturbed. The analysis suggested that static divergence contributed to the wing's resistance to return to an undisturbed state.

The study conjectures that relatively minor maneuvering differences during the first two high dihedral events may have prevented the aircraft from becoming highly unstable as occurred during the third high dihedral event.

Appendix D.1.c (Volume III)
Center-of-Pressure Location Calculated Using VSAERO for Various Dihedral Displacements and Angles of Attack
The purpose of this study was to explore the possibility that forward shifts in the center of pressure in the outboard sections of the HP03 vehicle could have contributed to the sustained high dihedral that led to the Helios instability. An analysis of the HP01 configuration was performed using the subsonic panel code VSAERO. The HP01 configuration was selected for this analysis because its geometry data was readily available from ARC, and the HP01 and HP03 aerodynamic configurations are sufficiently similar.

VSAERO panel models with various dihedral displacements were created from the original AV geometry. These geometries were analyzed at several angles of attack. The behavior of the sectional centers-of-pressure was examined as functions of dihedral and angle of attack. For comparison purposes, a rigid NASTRAN flat plate doublet-lattice linear aerodynamic model was created for the HP01 configuration and compared with some of the VSAERO results. For the conditions examined, the analyses indicated the following:

- Sectional centers of pressure are not functions of dihedral;

- Sectional centers of pressure move aft with increasing angle of attack;

- Sectional centers of pressure predicted using VSAERO are forward of those predicted by the NASTRAN doublet lattice code;

- VSAERO predicted the sectional centers of pressure near the wing tip to move forward; using the NASTRAN linear doublet lattice method the sectional centers of pressure near the wingtip were predicted to move aft.

Of these observations, the most significant is that VSAERO predicts the center of pressure to be forward of the quarter chord on the outboard wing panel. If the HP03 configuration is found to be very sensitive to key parameters like the elastic axis location, this observation could be important. Based on these analyses, there appears to be no purely aerodynamic coupling between center-of-pressure location and the dihedral deflection that could have led to the sustained high dihedral of the HP03 configuration. There was some sensitivity of center of pressure location to angle of attack, but it was not in a manner consistent with a static aeroelastic coupling that could have led to the high dihedral condition.

Appendix D.1.d (Volume III)
Identifying Possible Contributing Causes to the H03 Helios Mishap with
Static Aeroelastic Trim Analysis

This report describes the static aeroelastic trim analyses performed and highlights some of the sensitivities that the HP03 vehicle may be subject to when the aircraft is in trim or in steady longitudinal cruise conditions. Many of these sensitivities were brought about by the changes incorporated into the original aircraft (HP01) to configure it for the new, long-endurance-mission aircraft (HP03) involved in the mishap. The study used NASTRAN and linear doublet lattice aerodynamics to represent the aircraft. The parameters varied included: outboard wing panel incidence angle, fore-aft position of the hydrogen fuel tanks, wing-tip elevator deflection, and fore-aft relative positions of the wing spar and the wing center of pressure. Results indicated that:

- an increase in flight speed will result in higher wing dihedral;

- a forward shift in the center of pressure near the wing tip area with respect to the wing elastic axis tends to increase wing dihedral;

- the additive effects of outboard wing panel incidence angle, fore-aft position of the fuel tanks, and wing-tip elevator deflections affect wing dihedral.

Appendix D.1.e (Volume III)
Helios Mishap Data Review

This report provides the results of an investigation that analyzed flight data for the purpose of looking for features that may have been overlooked by those responsible for system design. The study independently verified important AV characterizations of the Helios flight control system, and provided some observations on the design and analysis methods used by AV.

Specifically, the author agreed with the majority of the AV analyses. However, he pointed out to AV engineering that the pitch-rate and airspeed-hold feedback paths cancelled each other out during high dihedral events leaving only a path involving the integral of pitch rate to command the elevator. Although such feedback is conventionally considered to be a suitable way to control phugoid dynamics, the analytical tools used by AV did not provide sufficient insight to optimize the feedback gains.

Comparing HP03-1 with HP03-2 flight data, the author recognized periods of higher normal load factor variations, which were interpreted as turbulence. These periods of higher-turbulence were well correlated with the close proximity to the qualitatively determined "shear lines." The author, however, was not persuaded that turbulence alone caused persistent high dihedral. Without delineating a continuous chain of causality, the author concluded, "The best course of action, in my opinion, is to make sure that the changes are made to the vehicle to assure that this level of dihedral (and resulting unstable phugoid) do not occur."

Power and Propulsion Working Group

Appendix D.2.a (Volume III)
Propulsion and Power Analysis: Focus on Electric Motor Performance
This report summarizes an analysis of the electric motor performance. The motor speed and approximate input power characteristics indicate that nothing in the Helios flight data suggests that the motors failed during the flight. A flight video clip, which shows that the aircraft propellers functioned properly as the mishap unfolded, supports this conclusion. There is further indication that the motors operated as expected until the instant of the mishap at which point the distortions in the flight data make the end result inconclusive. Typically, during climbing, the output torque of a given motor is below the limiting value. The video clip shows the Helios aircraft as still climbing when the mishap occurred. This infers that the motors were operating normally at the time of the mishap.

Appendix D.2.b (Volume III)
Helios Prototype Mishap Investigation: Lithium Primary Battery Data Review
This report summarizes an analysis of the lithium sulfur dioxide batteries that were used for the Helios aircraft. The primary batteries, Saft LO 26 SHX spiral D-sized 7.5 A-hr, provided the power for the take-off and landing. The Helios fuel cell configuration was flown with both main battery packs and an emergency battery pack located in Pods 2 and 3. The fuel cell pod has a connection to the emergency battery pack. At the time of the mishap both the solar arrays and the battery system were providing power to the aircraft. Due to the complexity of the system responses, it was difficult to single out any irregularities of power system performance during the last seconds of flight. Prior to the loss of signal, it appeared the three lithium-sulphur dioxide primary batteries were operating normally within their performance range.

Appendix D.2.c (Volume III)
Power and Propulsion
This report summarizes an analysis of the fuel cell system. The fuel cell power system was not in operation at the time of the mishap. Therefore, the fuel cell power system performance was not considered a cause or contributor to the mishap. Based on the reviews of the relative AV PDMs and the flight data, it appears that the solar arrays were performing within the expected values. At the end of the flight the performance starts to show irregularities when both solar array and battery system provided power to the aircraft. Due to a lack of visual evidence and inconclusive results when reviewing the test data, it was concluded that the irregularities associated with both power systems are due to structural breakout of the aircraft. After reviewing the flight data for all power and propulsion subsystems it was concluded that these systems were operating within the performance ranges established for the mission.

Structural Integrity Working Group

Appendix D.3.a (Volume III)
Helios Wing Structural Integrity Assessment
This report summarizes the results of an investigation to determine if the main wing spar near the right wing hydrogen fuel tank failed prior to the instability. During the study flight data, videographic records of the pre and post-failure events, and the failed wing spar hardware were assessed.

The analysis of the HP03-2 structural flight data concentrated on the strain gage data on the spar and trailing edge tube as well as selected accelerometer data from different spanwise locations near the leading edge of the vehicle. The visual inspection of the failed tubular spar at LaRC did not reveal any failure modes in the spar structure that suggested a progressive failure had occurred prior to the mishap. Most of the failures were judged to have been the result of a single high-load event, such as the wing's impact with the water or vehicle salvage from the ocean. The exception to this finding was that the spar failure site in the vicinity of the right hydrogen fuel tank mount did exhibit some failures in the outer Kevlar honeycomb layer that have progressive failure characteristics. However, the local carbon composite did not exhibit similar evidence of progressive failure. Furthermore, no in-flight data exists to suggest that this failure occurred prior to the initial vehicle break-up visible on the wing-tip video. Unfortunately, the lack of information following the break-up precludes drawing any conclusions as to the cause of this failure.

Based upon an examination of the wing-tip video, some potential failure-initiation scenarios were suggested in the report. A likely failure-initiating event was a local skin failure that allowed the wing cavity to become over pressurized, resulting in the wing skin peeling off of the rib flanges. One cause of a local skin failure was identified to be damage from the rib flange, especially if larger than normal relative motion between the foam leading edge and the ribs occurred during the periods of increased airspeed and dynamic pressure that occurred during the pitch oscillation events. Other causes include a failure of a spanwise seam or the presence of a skin repair. Another potential scenario is that the foam leading edge failed due to the high dynamic pressures. This would cause the rib leading edge structure, which is now not adequately stabilized, to buckle and collapse.

In summary, there was no evidence that suggests that there was prior structural failure(s) that would have contributed to the vehicle developing the higher-than-normal dihedral deflections that caused the unstable pitch oscillation and subsequent failure of the vehicle.

Appendix D.4.a (Volume II)
Numerical Simulations of Flow Regimes in the Lee of Kauai Including the Day of the Helios Breakup

Numerical simulations were carried out using a well-tested grid-point weather model (the "Clark-Hall" model) that has been widely used to study turbulence hazards to aviation. The model was used in an interactive nested-grid configuration where each successive nest (up to 5) had a shorter spacing between computational points with correspondingly smaller computational domain. The outer nest covered the entire Hawaiian Island chain and had horizontal grid spacing of 6 km. The innermost nest had a grid spacing of 167 m in the horizontal and 50 m in the vertical and was centered over the Kaulakahi Channel a little northeast of the mishap location.

Simulations were initialized with real data. In many of the simulations the National Weather Service balloon soundings (rawinsondes) launched at Lihue, Kauai, near the southeast coast, were assumed to be representative of the atmosphere over the entire computational domain. In some others the model was initialized using input from another weather model run by the UH for this purpose (see Appendix D.4.b). Simulations were made for the day of the mishap (26 June 03), for the day of the HP03-01 flight (7 June 03), and for one of the UH research flight days (8 September 03).

It was not possible to exactly replicate the atmospheric conditions at the time of the mishap using a weather model. However, for aspects of the model that could be compared at least qualitatively with observations (wake region, north and south shear lines, strength of winds, locations of turbulence on the UH 8 September flight day, etc.) the model appeared to produce a very credible simulation. This is particularly true given the surprising amount of uncertainty in the overall strength and direction of the trade winds as well as in winds at higher levels.

Principal conclusions from these simulations are:

- Shear line (or zone) strength and location are very sensitive to the trade wind direction and speed;

- Well defined shear zones were present on 26 June 03;

- The shear zones and the inferred amount of turbulence within them were much weaker on 7 June 03 than on 26 June 03;

- Turbulence is concentrated in the shear zones; away from the shear zones, both within the wake region and over the open ocean, turbulence is significantly less;

- In many of the simulations the shear lines had a marked northward tilt with height, thus the use of sea-state discontinuities may not be a reliable means to estimate the position of the shear lines aloft;

46

- The shear zones are unstable and transitory, therefore the sea state is only an indication of the mean location of the shear lines (or zones) at the surface;

- The shear zones exhibit breakdown into smaller-scale gust features; in the simulations of 26 June03, horizontal wavelike structures appeared with a slightly preferred NW-SE orientation;

- Within the shear zones a lower bound estimate of turbulence 2σ vertical winds is about 1 m/sec;

- The effects of solar heating are to increase the turbulence levels in the lowest 1500 - 2000 feet of the atmosphere; locations of shear lines and intensity of turbulence within them above this altitude are little affected.

Appendix D.4.b (Volume II)
Meteorological Investigation of the Helios Mishap: Modeling and Observations
This report summarizes:

- overall weather situation on 26 June 03;

- results of simulation of the airflow in the vicinity of Kauai on 26 June 03 using a well-tested weather model (NCAR/Penn State Mesoscale Model Version 5);

- special observations of airflow and turbulence in the vicinity of HP03-2's flight trajectory made from an instrumented light aircraft under trade wind conditions several weeks after the mishap.

The airflow in the vicinity of the Hawaiian Islands on 26 June 03 was very similar to the climatological mean for this time of year. In particular, available observations and indications from satellite of the regions of inferred weaker winds (wake regions) to the lee of the islands indicated the trade winds were from the east-northeast (ENE) over most of the island chain, veering to easterly in the vicinity of Kauai, which is the farthest west of the main islands. Above about 6000 feet altitude the trade winds weakened and winds above about 18,000 feet were from the southwest (SW) quadrant, approaching 50 knots near 40,000 feet.

For days with trade-wind flow similar to that observed on 26 June 03, both the observations from the specially instrumented aircraft and the model simulations (as well as the model simulations reported in Appendix D.4.a) are consistent in showing the presence of a wake region to the lee of Kauai produced by the blocking effect of Kauai's central mountains. The central mountains rise to 5,000 feet, and are therefore sufficient to strongly perturb the trade-wind flow; a large fraction of this flow is forced to go around the central mountains to the north or south, rather than over them. This leads to both a local enhancement to the trade-wind flow north and south of Kauai, and perturbations ("mountain waves") in the flow that traverses the mountains and flows westward above the wake region. It should be noted that a major thrust of the UH modeling effort was directed toward providing initial and time-dependent lateral boundary conditions for

47

very high-resolution simulations carried out by NCAR for 26 June 03 and reported on in Appendix D.4.a.

The thrust of the aircraft study was to collect measurements in this wake region west of PMRF over the Kaulakahi Channel and the regions of horizontal and vertical wind shear that bound it laterally and vertically. Note that PMRF is nearly always in this wake region, which originates on the lee (west) side of the central mountains east of PMRF. Flights were made on 6 days; all flight days had trade-wind flow from directions between NE and E to ESE. Surface evidence of the shear lines, based on sea state, were observed on all flights, but the positions of the shear lines were found to be dependent on the trade wind direction and were more-or-less aligned along this direction. Thus, on days with trades more easterly, the shear lines over the channel were farther north than on days with trades from the NE. The surface position of the shear lines was not necessarily straight, and sometimes was discontinuous, suggestive of eddies, and was also not steady state. Nor can we generalize to say that the shear lines always got closer together farther offshore. The shear lines (or zones) were evident in the winds measured by the UH airplane to above 3000 feet altitude, but the horizontal wind shear was strongest at 2000 feet and below. The shear lines were not apparent in measured winds at 6000 feet and above. The shear lines tended to be nearly vertical in orientation (as judged by comparing location of measured horizontal shears with surface observation of gradient in sea state) or, particularly in the case of the northern shear line, tilted slightly to the North. Near 500 feet altitude, turbulence (as gauged from UH aircraft vertical accelerations) in the wake region was only slightly reduced from that near the shear lines, but at higher altitudes, turbulence was distinctly concentrated along the shear lines. However, flights at 6000 feet and above had very little turbulence. The wake region above 500 feet altitude was more turbulent than the air on the open-ocean side of the shear lines.

Analysis of sun glint, noted from a polar-orbiting satellite pass over Kauai about 45 min after the mishap, indicated that the wake region (weaker winds) to the lee of Kauai extended westward at least 100 km from PMRF, with the implication that some vestiges of the north and south shear lines extended at least that far downstream at the time of the mishap on 26 June 03. Vertically integrated water vapor measured on the same pass detected an east-west trail of higher water vapor extending westward from just north of PMRF across the location of the Helios breakup.

Appendix D.4.c (Volume II)
Helios Report - Environmental
This report discusses the following:

- Kauai Weather: Description of weather patterns that impact flight weather, especially the trade winds and local sea and land breezes;

- Meteorological Instrumentation: Balloons, sodars, towers, and radar;

- Weather hazards and constraints: List of 12 specific atmospheric criteria and 6 general guidelines on phenomena to avoid;

- Flight day weather support: Description of weather support to operations, including forecasts provided to flight crew and the meteorologists' concerns about weather conditions on 6 June 03;

- Shear Lines: Discussion of the causes and behavior of shear lines; tools available to analyze shear lines and low level winds; and a comparison of 6 June 03 low level wind conditions as compared to the other PMRF flight days;

- Situational Awareness: Numerous cautionary factors known before flight are listed. When assessed together, they should have peaked the flight team into a much greater *situational awareness,* and sparked a very elevated pre-flight sensitivity to unusual aircraft behavior and a tough discussion regarding what immediate actions should follow if any anomalies were observed; for a list of the cautionary factors, see Section 3.2.2 of the Events and Causal Factor Tree in Section 9.

- Recommendations:

    1) Little was known before Helios 03-2 about the dynamics of shear line/ mountain wave/ sea breeze interactions. Our very brief investigation with NCAR and UH indicates the shear lines are very dynamic. However, there's still much more to learn before we know what is possible even statistically, much less through rigorous analysis. In the wake of an accident, common reaction is, if possible, to redesign the vehicle or system to eliminate the vulnerability that caused the mishap. In the case of Helios, the obvious mitigation is to strengthen the main spar and increase its restorative tendency to counteract increasing dihedral. The purposes are to: 1) Decrease the risk of mission failure; and 2) Increase the range of atmospheric conditions that can be flown in and thus increase flight/mission opportunities. However, improving the understanding of the atmosphere can also do both. Given that users will almost always attempt to operate close to the constraint boundaries and not miss opportunities, the obvious conclusion is to work both issues—design and the atmosphere.

    2) Avoid shear lines, including the assumption of vertical slope, at least until their structure and behavior are better understood and flight dynamics models are improved to better assess atmospheric impacts on stability and control.

    3) Instrument aircraft to better understand what the aircraft is encountering and has encountered. Purpose: Improve preflight accuracy of aircraft performance tools, and post flight ability to validate these tools and the atmospheric models input into them, by reducing uncertainty introduced by the atmosphere.

    4) Document weather constraints or conditions, which would prompt immediate discussion regarding Return to Base.

    5) Do not transfer aircraft from one team to another during more precarious portions of flight such as climb out.

49

6) Perform thorough investigation of weather dynamics in flight test area whether Kauai, Afghanistan, etc., or only fly within areas/ranges where mesoscale/microscale weather phenomena and their characteristics are well documented.

7) Engage advanced degree atmospheric physics experts, from at least NASA and contractor, in entire development and testing process from womb to tomb, including aircraft design, concept of operations formulation, modeling, and post flight analyses.

8) Weather Infrastructure:

- Diagnose and fix problem with Sodar 2000 or install similar instrument in order to resume obtaining high-resolution time measurements of horizontal wind and vertical velocity variance to altitudes reaching at least 2000 feet.

- Establish procedures to obtain additional radar products, for example via FTP from Doppler weather radar near Port Allen, Kauai, if it appears they would be useful for identifying shear line locations or characteristics.

- Survey remainder of Kauai to identify sources of and methods to access additional weather data, especially along the north and south coasts of Kauai and on Niihau.

- Dual and ideally triple Lidar to provide sector scans of wind along flight profile.

Appendix D.4.d (Volume II)
Low Altitude Winds Experienced on Solar Powered Aircraft Flight Days
Wind concerns for test flights from PMRF include:

- Runway wind speed and direction including vertical motion variance immediately above the runway for takeoff and recovery.

- Strong wind shears aloft and other meteorological indicators of turbulence, including location of northern and southern shear lines.

- Wind speed vs. the airplane true airspeed as a function of altitude for navigation.

- Wind speeds and direction as a function of altitude to calculate trajectory of mishap debris to ensure it impacts within acceptable area.

This report primarily focuses on Trade Winds which would most strongly influence #2 above. Flight operations personnel characterized the 26 Jun 03 trade winds as 'unusual'—blowing from more south of east than normal and stronger than normal. The DFRC Meteorologist compared lower level wind profiles from PMRF balloon releases and concluded the following about 26 June 03 PMRF winds as compared to those on 16 other PMRF flight days:

- Winds at 5000 feet were approximately 18 knots, equal to or less than winds on three other flight days.

- The maximum low-level wind speed was 18 knots compared to winds of 16, 17, 20, 23, and 24 knots on five other PMRF flight days.

- Vector wind shear near the mishap was .021 per second, compared with seven other flight days >.02 per second.

- 5000 feet wind directions were around 104 degrees, similar to seven other flight days, which ranged from 095 to 115 degrees.

A similar analysis of winds at ~ 350 feet, 3500 feet, and 5000 feet from Lihue balloons at 8 hours before and 4 hours after the mishap concluded:

- Wind speeds on 26 June 03 were in lower 20 - 40 percentile of the 17 PMRF flights; one exception: 5000 feet wind 3 hours after the mishap ranked at 62 %. These lower level wind speeds were slightly below or equal to normal wind speeds for June.

- Wind directions were typical of other flights. Exception: 5000 feet wind was more southerly (095°) than average (075°).

Overall 26 June 03 conditions were close to climatological averages and within the envelope of wind speeds and directions experienced on previous solar powered aircraft flights at PMRF. Compared to other PMRF flight days, the 850 mb wind speed, and the maximum low altitude speed and shear were all stronger than the median day conditions, and 26 June 03 would rank near the 75th percentile relative to previous PMRF flight wind conditions. In contrast, the upstream forcing wind speeds at Lihue were less than the median and would rank near the 30th percentile.

Winds derived along the Helios flight path exhibit significantly higher shear values than the rawinsondes at PMRF, particularly in the layers 700 – 1400 feet and 2700 – 3000 feet MSL. Interpreting the wind changes in these layers as vertical wind shear, that is as the rate of horizontal wind vector change with altitude, and adjusting for variations due to Helios' horizontal motion and altitude change, appears to yield inordinately high values of shear. However, the author cautions, 'it is important to confirm (or reject) the Helios derived winds via inspection of airplane data for errors'. It is possible that the large shear values the author notes are due to lack of numerical precision in GPS altitude of Helios or from the assumption that wind vector changes along the flight path have no contribution from horizontal shear or unsteadiness of the flow.

Appendix D.4.E (Volume II)
Weather Perspective on Helios Mishap
This report discusses some perspectives of the Helios MIB meteorologists concerning the contribution of atmospheric disturbances or "turbulence" to the Helios mishap. Major points are as follows.

- From interactions with aerodynamics and control experts who were Board members or advisors to it, the pitch instability was known to be a characteristic of a wing-tip deflection (dihedral) of more than about 30 feet for HP03. This high dihedral had to have been initiated by atmospheric flow disturbances that caused larger lift on the outboard wing panels than near the centerline for sufficient duration for the airplane to respond.

- Based on conclusions by the aerodynamics experts, the sensitivity of the HP03 to such gusts was considerably larger than that of HP01.

- The flight of HP03-1 on 7 June 2003 occurred in exceptionally smooth air. Numerical simulations of the atmospheric flow and turbulence on this day emphatically supported this conclusion, made by the crew on 7 June.

- The flight of HP03-2 was made in atmospheric conditions much more typical of previous successful high altitude flights from PMRF. Trade winds were average to somewhat above average in strength relative to previous flights, and from a slightly more easterly direction than the typical east-northeast. Winds were clearly within the envelope of previous flight days. The wind-shear lines (which form a boundary between the light winds extending offshore from the western shore of Kauai (including PMRF), where there is shadowing by the island terrain), and the trade winds (which are deflected around the island by this terrain), were well marked. From numerical simulations and actual instrumented aircraft flights, all conducted as part of the MIB investigation, these shear lines are confirmed to be regions of enhanced turbulence, especially between about 1,500 and 4,000-foot altitude. The Helios was close to these shear lines at the time of the mishap.

- Detailed numerical simulations showed the existence of disturbances of the vertical component of air motion in the northern of the two shear lines that had potential for giving rise to development of high dihedral in HP03-2. This was due to their form and orientation. A further conclusion from the simulations is that horizontal shear of the vertical component of air motion on 26 June equaled or exceeded .01/sec 2-3 percent of the time in the northern shear line.

- Meteorological expertise is available (within and outside NASA) that should be tapped during both vehicle design and the development of operational procedures to help anticipate weather hazards that might be encounter in future HALE operations at other locations.

Section 9
Events and Causal Factor Tree Analysis

The MIB as well as the AV independent investigation team attempted to consider all possible causes of the mishap. The AV independent investigation team identified possible causes and arranged them into two cause-and-effect trees: one for the aircraft and its control system; and the other for crew actions. In parallel, the MIB also created an Events and Causal Factor Tree. Although the AV cause-and-effect trees and the MIB Events and Causal Factor Tree were created independently they contained almost identical information and potential causal effects.

Figure 9.1 provides a legend to better follow the MIB Events and Causal Factor Tree shown in Figure 9.2. Color-coding correlates to the identification of root causes, contributing factors, significant observations, and findings as presented in Section 10. For example, root causes can be identified by a red colored event symbol (box with circle below). Gates, representing intermediate or summary conditions between potential causal factors and the top level event of "uncontrolled motion about any axis" are colored to reflect the most significant of the four categories of events that define it. Increasing event significance is ordered from right to left in the legend, with root causes having the greatest significance relative to the mishap.

Table 9.1 shows the Events and Causal Factor Tree in WBS format; "AND gates" are shown in **bold** and "OR gates" are shown in *italic*. Faults or events that were determined to be "true" are denoted as (T) and faults or events that were determined to be "false" are denoted as (F). Further details describing each element of the Events and Causal Factor Tree are provided below and in many cases, a reference that presents additional evidence as to the path being declared "True" or "False" is provided.

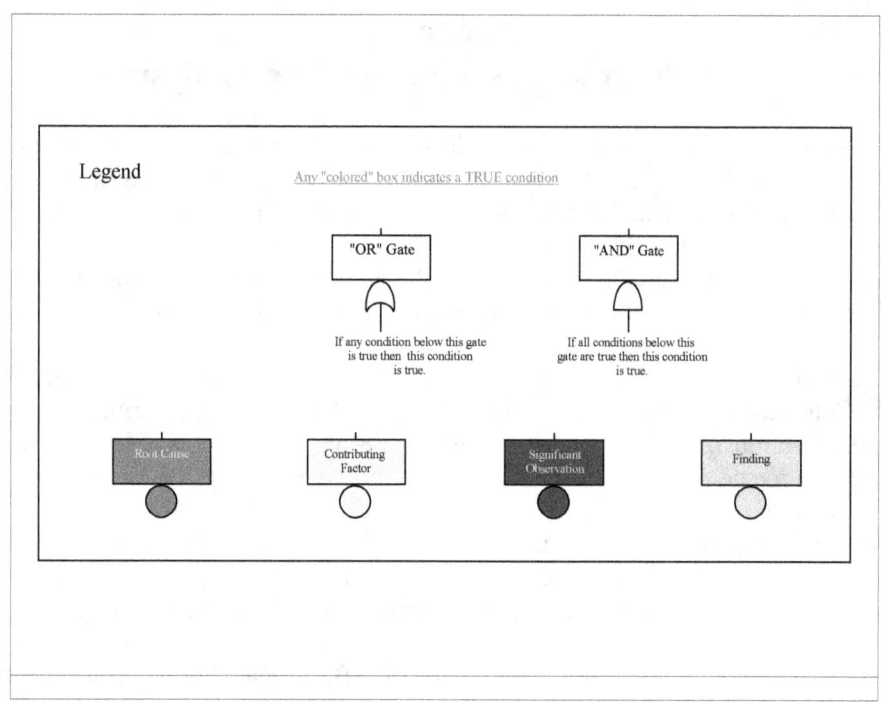

Figure 9.1 – Legend for the MIB Events and Causal Factor Tree

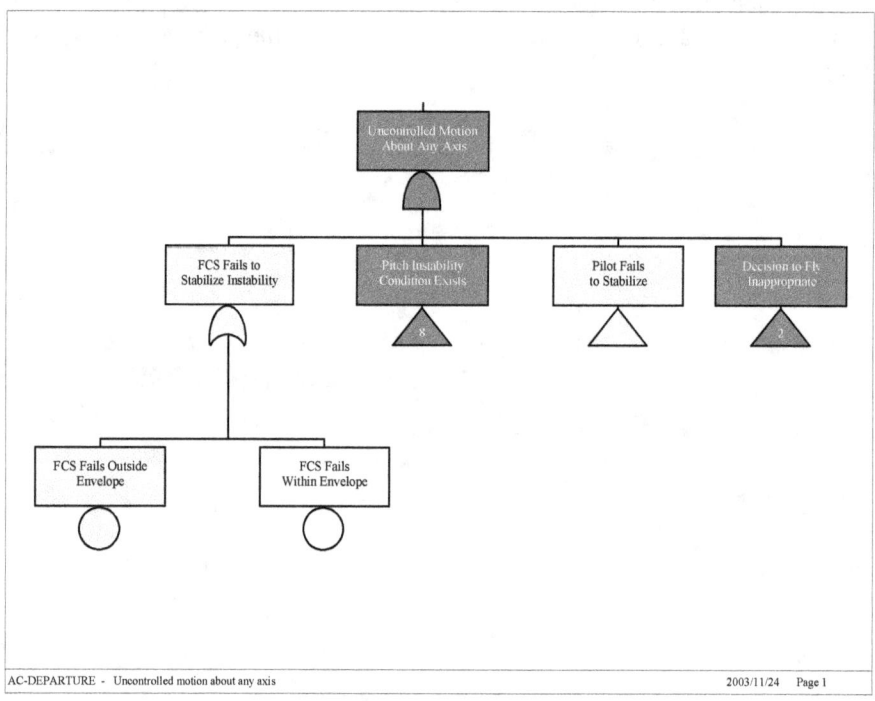

Figure 9.2 (1 of 7) – MIB Events and Causal Factor Tree

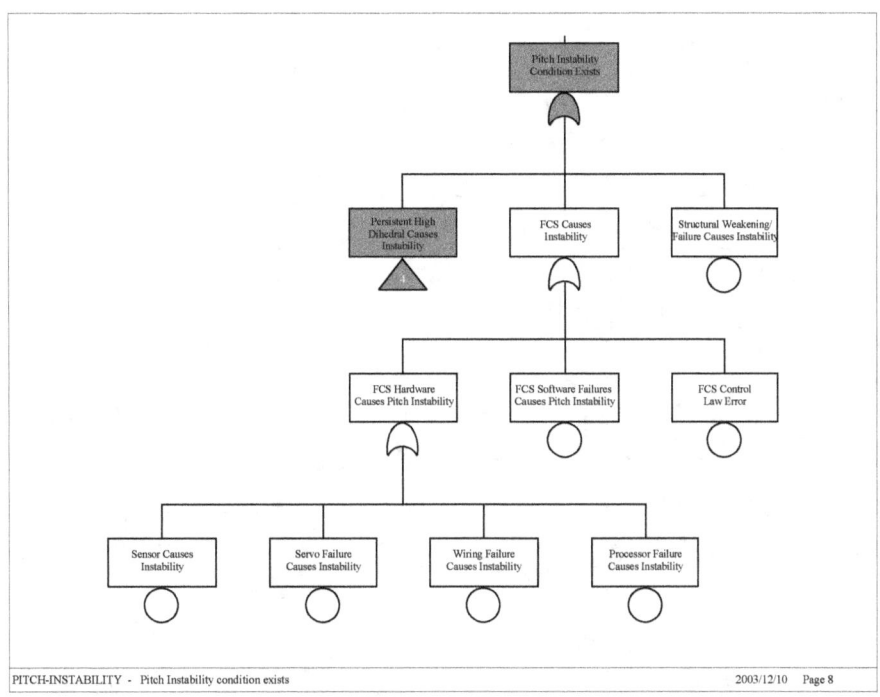

Figure 9.2 (2 of 7) – MIB Events and Causal Factor Tree

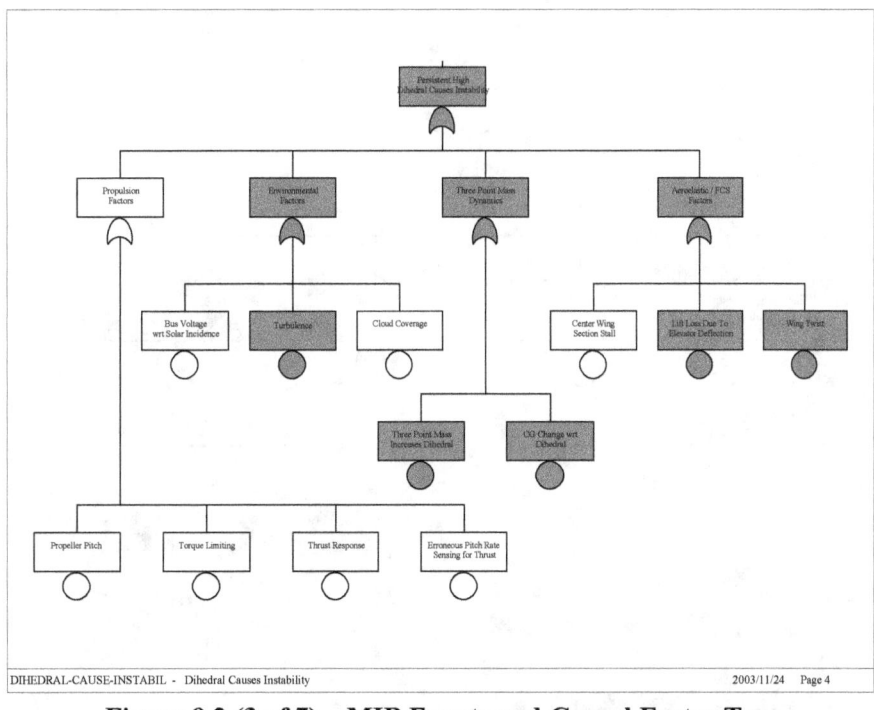

Figure 9.2 (3 of 7) – MIB Events and Causal Factor Tree

55

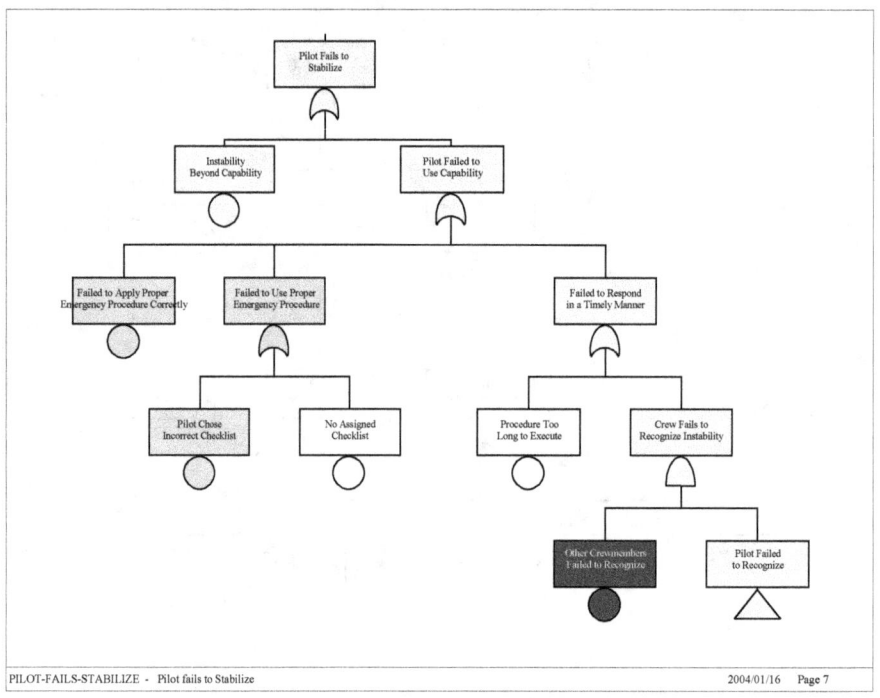

Figure 9.2 (4 of 7) – MIB Events and Causal Factor Tree

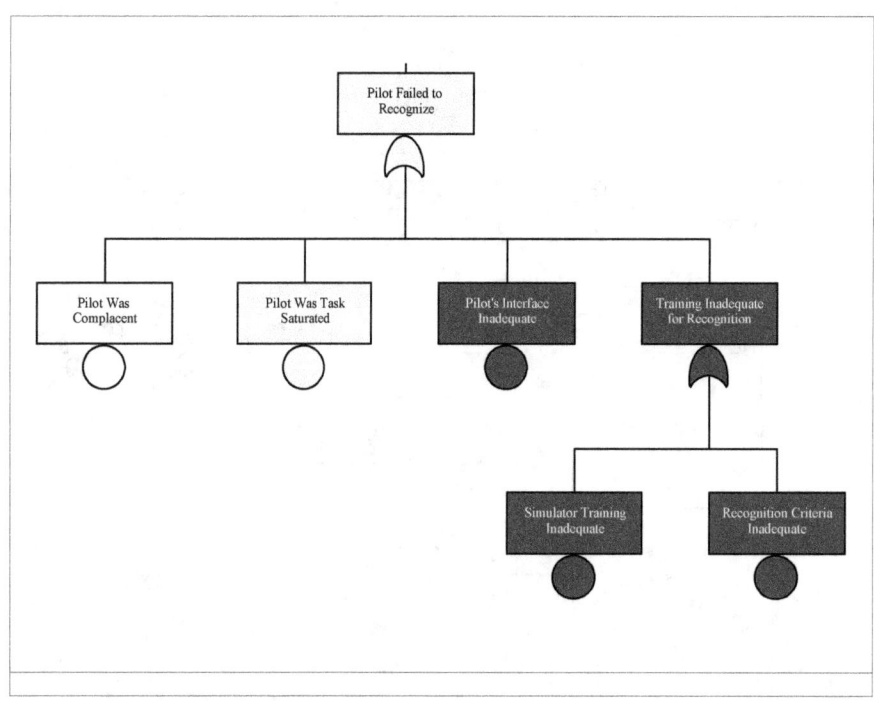

Figure 9.2 (5 of 7) – MIB Events and Causal Factor Tree

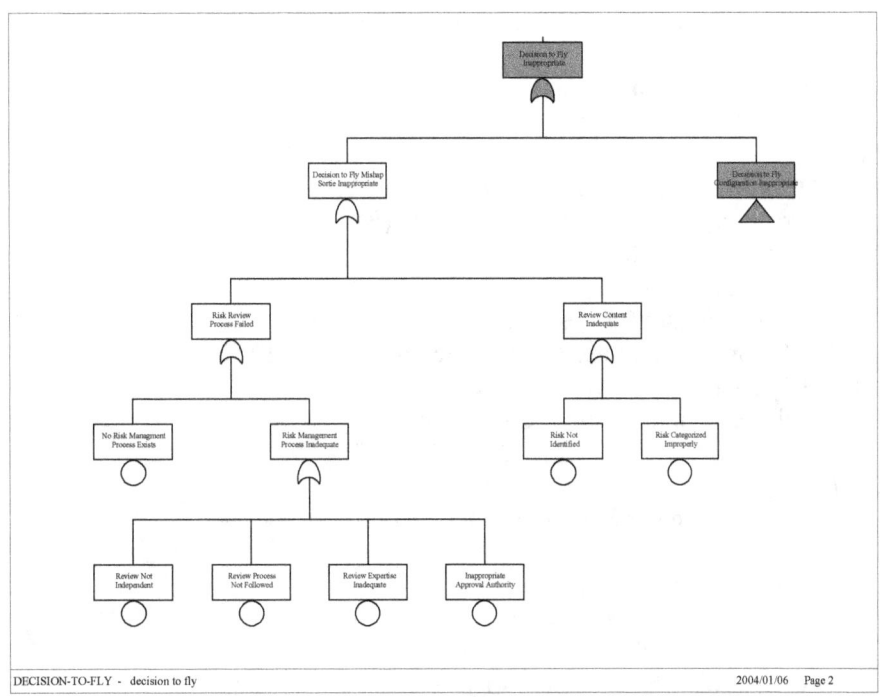

Figure 9.2 (6 of 7) – MIB Events and Causal Factor Tree

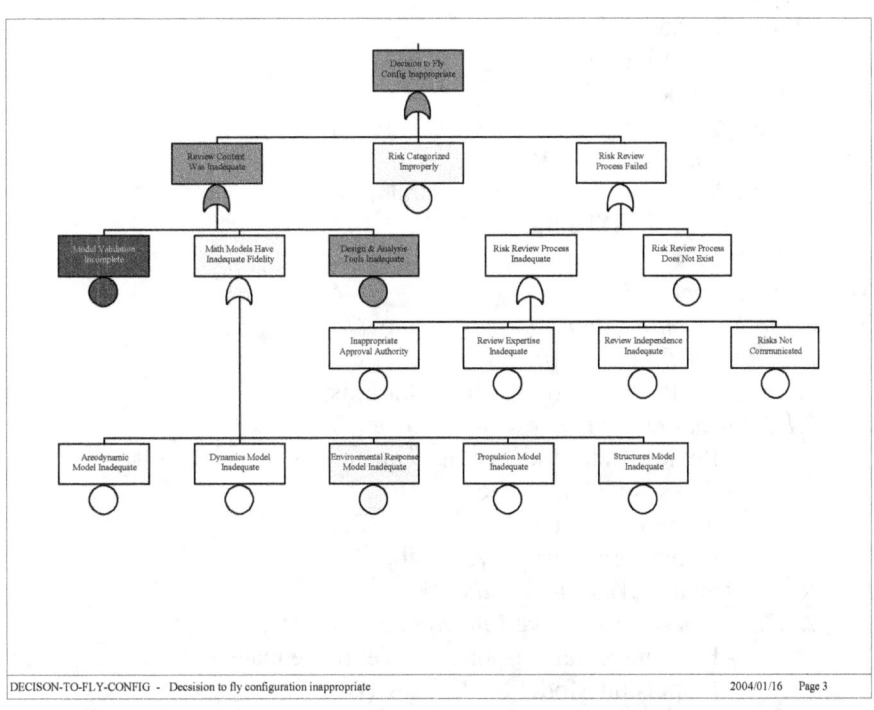

Figure 9.12 (7 of 7) – MIB Events and Causal Factor Tree

57

Table 9.1 – MIB Events and Causal Factor Tree

Uncontrolled Motion about any Axis (T)
1.0 Pitch Instability Condition Exists (T)
   1.1 FCS Causes Instability (F)
      1.1.1 FCS Hardware Causes Pitch Instability (F)
         - Processor Failure Causes Instability (F)
         - Sensor Causes Instability (F)
         - Servo Failure Causes Instability (F)
         - Wiring Failure Causes Instability (F)
      1.1.2 FCS Software Failures Causes Pitch Instability (F)
      1.1.3 FCS Control Law Error (F)
   1.2 Structural Weakening / Failure Causes Instability (F)
   1.3 Persistent High Dihedral Causes Instability (T)
      1.3.1 Environment Factors (T)
         - Bus Voltage wrt Solar Incidence (F)
         - Turbulence (T)
         - Cloud Coverage (F)
      1.3.2 Aeroelastic / FCS Factors (T)
         - Wing Twist (T)
         - Center Wing Section Stall (F)
         - Lift Loss Due to Elevator Deflection (T)
      1.3.3 Propulsion Factors (F)
         - Propeller Pitch (F)
         - Torque Limiting (F)
         - Thrust Response (F)
         - Erroneous Pitch Rate Sensing for Thrust (F)
      1.3.4 Three Point Mass Dynamics (T)
         - Three Point Mass Increases Dihedral (T)
         - CG Change wrt Dihedral (T)

2.0 Decision to Fly Inappropriate (T)
   2.1 Decision to Fly Configuration Inappropriate (T)
      2.1.1 Risk Review Process Failed (F)
         2.1.1.1 Risk Review Process Does Not Exist (F)
         2.1.1.2 Risk Review Process Inadequate (F)
           - Review Independence Inadequate (F)
           - Risks Not Communicated (F)
           - Review Expertise Inadequate (F)
           - Inappropriate Approval Authority (F)
      2.1.2 Review Content Was Inadequate (F)
         2.1.2.1 Math Models Have Inadequate Fidelity (F)
           - Environmental Response Model Inadequate (F)
           - Propulsion Model Inadequate (F)
           - Dynamics Model Inadequate (F)
           - Aerodynamic Model Inadequate (F)

        - Structures Model Inadequate (F)

        2.1.2.2 Design and Analysis Tools Inadequate (T)

        2.1.2.3 Model Validation Incomplete (T)

    2.1.3 Risk Categorized Improperly (T)

2.2 Decision to Fly Mishap Sortie Inappropriate (F)

    2.2.1 Risk Review Process Failed (F)

        2.2.1.1 No Risk Management Exists (F)

        2.2.1.2 Risk Management Process Inadequate (F)

            - Review Not Independent (F)

            - Risk Process Not Followed (F)

            - Review Expertise Inadequate (F)

            - Inappropriate Approval Authority (F)

    2.2.2 Review Content Inadequate (F)

        2.2.2.1 Risk Not Identified (F)

        2.2.2.2 Risk Categorized Improperly (F)

3.0 Pilot Fails to Stabilize (T)

    3.1 Instability Beyond Capability (T)

    3.2 Pilot Failed to Use Capability (T)

    3.2.1 Failed to Use Proper Emergency Procedure (T)

        3.2.1.1 Pilot Chose Incorrect Checklist (T)

        3.2.1.2 No Assigned Checklist (F)

    3.2.2 Failed to Respond in a Timely Manner (T)

        3.2.2.1 Crew Fails To Recognize Instability (T)

            3.2.2.1.1 Pilot Failed to Recognize (T)

                3.2.2.1.1.1 Training Was Inadequate for Recognition (T)

                    - Simulation Training Inadequate (T)

                    - Recognition Criteria Inadequate (T)

                3.2.2.1.1.2 Pilot Was Task Saturated (T)

                3.2.2.1.1.3 Pilot Was Complacent (F)

                3.2.2.1.1.4 Pilot's Interface Inadequate (T)

            3.2.2.1.2 Other Crewmembers Failed to Recognize (T)

        3.2.2.2 Procedure too Long to Execute (F)

    3.2.3 Failed to Apply Proper Emergency Procedure Correctly (T)

4.0 FCS Fails to Stabilize Instability (T)

    4.1 FCS Fails Outside Envelope (T)

    4.2 FCS Fails Within Envelope (F)

Detailed Discussion of the MIB Events and Causal Factor Tree

Uncontrolled Motion about any Axis (TRUE)
The loss of control of the HP03-2 resulted in a nose down dive, which allowed the aircraft to exceed the maximum allowable speed by over a factor of two. The resulting excessive dynamic pressure caused the foam leading edge structure and the solar cell wing covers at the wing outer panels to separate from the aircraft. This separation destroyed the aircraft's ability to generate lift and sustain flight. Four potential conditions had to exist for the uncontrolled motion about any axis to occur. These conditions were: 1) a pitch instability occurred; 2) an inappropriate decision to fly was made; 3) the pilot failed to stabilize the aircraft; and 4) the flight control system failed to stabilize the aircraft. It was determined that all of these caused the uncontrolled motion about the pitch axis. It was also determined that none of these caused uncontrolled motion about the lateral or directional axis. Only the uncontrolled motion about the pitch axis is discussed below since this is the relevant axis.

1.0 Pitch Instability Condition Exists (TRUE)
The nose low steep dive (approaching 90°) was primarily the consequence of an unstable pitch oscillation, a known flight dynamic characteristic of the Helios that had been predicted, observed, and controlled in prior flights with predecessor configurations. The team, however, did not predict the high rate of divergence of this instability. Three potential causes of the pitch instability condition were explored. These included: the flight control system (FCS), structural weakening or failure, and a persistent high dihedral. It was determined that a persistent high dihedral was the primary cause of the pitch instability.

1.1 FCS Causes Instability (FALSE)
Three potential ways for the FCS to cause the pitch instability were explored. These included hardware failures, software failures, or control law errors. It was determined that the FCS did not cause the pitch instability.

1.1.1 FCS Hardware Causes Pitch Instability (FALSE)
Four potential hardware failure modes were explored during the investigation. The investigation found no evidence that indicated failures in the FCS processor, sensors, servos, or wiring had occurred. "AeroVironment performs a routine examination of downlinked telemetry after every flight. Every data record is examined for evidence of anomalous behavior. This process is the key to assurance that the backup systems all perform as expected and that any unusual event is evaluated. The only anomaly, noted in Document #5 found in Volume III, Appendix E, related to the flight control system was that the magnetic compass was off by 20 degrees when the system was powered up. This was determined to be due to significant magnetic variation on the transient ramp where the aircraft is mated prior to flight. The lateral control mode in use at the time of the mishap did not use magnetic heading." (Excerpt taken from Document #25 found in Volume III, Appendix E, page 36)

1.1.2 FCS Software Failure Causes Pitch Instability (FALSE)
The aircraft control loops and gains are implemented in software. Concerns about control system functions are mostly software issues. Correct software function was

verified by constructing a check of the FCS by using sensor signals, as downlinked from the aircraft, to reconstruct the FCS computer outputs. The calculated outputs are then compared to the downlinked data. Document #22 found in Volume III, Appendix E summarizes a check of the longitudinal control loop for flight HP03-2. The elevator control signal was recreated from the downlink data during the final one and a half minutes of flight HP03-2, including the divergent pitch oscillation, so that it could be compared to the actual elevator command to verify the correct operation of the control system. The command and sensor data were imported into an Excel spreadsheet to perform the investigation. Each component of the elevator control signal was recreated from this data and then summed to give the total elevator signal. This analysis indicated that the control system was in fact working correctly for the entire period.

The commanded elevator control deflections were exactly as predicted by the spreadsheet recreation of the FCS computer software. The elevator commands were appropriate to damp the oscillation. After the oscillation went unstable, the elevator control commands hit two limits. The first was the airspeed integrator limit, and the second was a limit in the pitch rate sensor itself. Document #23 found in Volume III, Appendix E summarizes the check of the lateral control loop for flight HP03-2. The steering motor command signals were recreated from the downlink data and compared to the actual steering motor commands. It was found that the actual motor commands behaved as programmed, in other words, the control system did what it was designed to do.

1.1.3 FCS Control Law Error (FALSE)

"Document #9 found in Volume III, Appendix E is a 125-page report that details the analytical method used to determine flight control system gains for the Helios aircraft. It presents the analysis used to set gains for the first flight of 2003, comparison of flight data to the expected system behavior, changes made between first and second flights to improve the system, and the analysis used to set gains for the second flight. This extensive review, which included a new analysis, did not discover any errors in the process used to set and evaluate the system gains within the system limitations noted above." (Excerpt taken from Document #25 found in Volume III, Appendix E, page 35)

"An end-to-end gain check is conducted each time the flight control computer software is changed. This procedure, summarized in Document #8 found in Volume III, Appendix E, was conducted as usual prior to Flight H03-2. It verified that the correct gains had been loaded. The investigations noted in the previous section also serve to verify that the correct gains were functional for all control loops relevant to the mishap." (Excerpt taken from Document #25 found in Volume III, Appendix E, page 37)

1.2 Structural Weakening / Failure Causes Instability (FALSE)

The results of a structural integrity assessment of the damaged HP03-2 wing spar are summarized in Volume III, Appendix D.3.a. The scope of this investigation was centered on an assessment of the flight data, the videographic records (Document #7 found in

Volume III, Appendix E) of the pre- and post-failure events, and the failed wing spar hardware. The visual inspection of the failed tubular spar (see the photographs provided in Document C.2, Appendix C, Volume II) at LaRC did not reveal any failure modes in the spar structure that suggested a progressive failure had occurred prior to the mishap. Most of the failures were judged to have been the result of a single high-load event, such as the wing's impact with the water. The exception to this finding was that the spar failure site in the vicinity of the right hydrogen-air fuel tank mount did exhibit some failures in the outer honeycomb layer that have progressive failure characteristics. However, the local carbon composite did not exhibit similar evidence of progressive failure. The study concluded that there was no evidence that suggests that failure of primary structural prior to the mishap contributed to the vehicle developing the higher-than-normal dihedral deflections.

1.3 Persistent High Dihedral Causes Instability (TRUE)

It was known that the aircraft with a persistent high dihedral above 30 feet would be unstable. Four potential ways for a persistent high dihedral to exist were explored. It was determined that a combination of environmental, aeroelastic, FCS, and point mass effects initiated and/or caused the persistent high dihedral, which in turn caused the pitch instability. It was also determined that the propulsion systems did not contribute to sustaining the persistent high dihedral observed.

1.3.1 Environment Factors (TRUE)

Three potential ways for the environment to initiate or cause the persistent high dihedral were explored. It was determined that neither bus voltage with respect to solar incidence nor cloud coverage initiated or caused the observed persistent high dihedral. However, turbulence was determined to have initiated the high wing dihedral, but was not necessary to maintain it.

- Bus Voltage wrt Solar Incidence (FALSE)

A theory that power fluctuations resulting from the changing sun incidence angle on the wing solar cell array caused the aircraft pitching motions was considered and analyzed. Reviews of test data and Documents #18 and #26 found in Volume III, Appendix E provide evidence that changes in solar incidence did not cause the mishap. The sun at 10:30am local time was essentially normal to the wing solar cell array. During the last few seconds of the flight, the motor performance starts to show irregularities when both the solar array and battery system are providing power to the aircraft. The analysis indicated that the power fluctuations did become large during the final, extreme pitch oscillations of the aircraft. However these fluctuations did not influence the unstable pitching motions of the aircraft.

- Turbulence (TRUE)

Post mishap analysis revealed that spanwise lift redistribution (from the center of the aircraft to the outboard wing panels) was very sensitive to small amplitude gusts. It has been the experience of the Helios type aircraft that its dihedral varies as local airflow varies along the wingspan. The ASWing code was used to estimate the spanwise lift distribution sensitivity for variations in vertical gust velocity.

Figure 9.3 provides a comparison of the wing spanwise lift distribution for the HP01 and HP03-2 aircraft configurations with a superimposed 0.5 ft/sec gust. For this study, it was assumed that the aircraft encountered a gust exhibiting a 0.5 ft/sec downward flow at the centerline and 0.5 ft/sec upward flow at the wing tips. Although the probability of the aircraft encountering such a gust is very low, the intent of the analysis was to show the sensitivity of the aircraft to gusts, even small gusts. The HP03-2 configuration has noticeably more lift outboard and less lift inboard. The increased lift outboard would be expected to lead to considerably increased dihedral. It is important to remember that the modeled sensitivity to local variations in sectional airflow is primarily a consequence of the aircraft's reactions to increasing dihedral, trailing edge up elevator, and wing tip leading edge twisting up, not the first-order effects of airflow change. The airflow variation is only the transitory force. The elevator response and the change in twist are what make the wing slow to return to nominal dihedral conditions. Document #26 found in Volume III, Appendix E provides a detailed discussion of this finding. In the absence of a gust that flattens the wing (for example, a momentary gust having relative downward air motion on the tips and upward near the airplane centerline), these aerodynamic effects slowed the return to nominal dihedral.

Although general weather conditions seemed to be within the bounds present during previous PMRF flights, HP03-2's faster air speed and unchanged climb rate led to a different flight profile with greater exposure to the turbulence. As a result, HP03-2 flew a greater distance at low altitudes and thus was exposed to the turbulent patterns induced by Kauai for a longer period of time. Also, the HP03-2 encountered the shear lines at a lower altitude than previously, closer to the shear lines' regions of maximum turbulence.

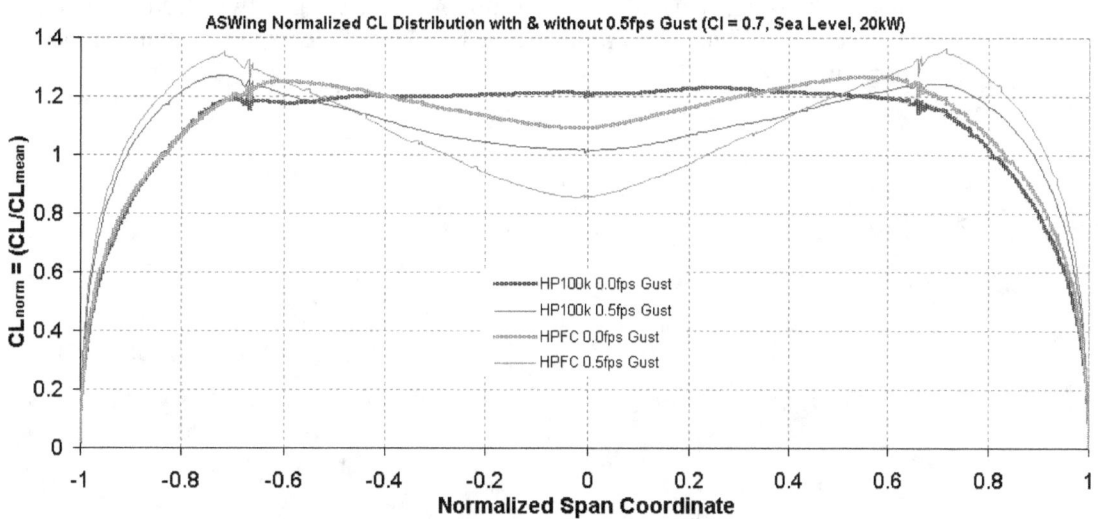

Figure 9.3 – Variation Spanwise Lift Distribution with Gust Velocity

- Cloud Coverage (FALSE)
An analysis was performed to determine if low clouds with well-defined edges resulting in sudden changes in solar power could cause the unstable pitch events. This analysis and the results are summarized in Document #17 found in Volume III, Appendix E. The study showed that the power variations, even though they were large, did not excite oscillatory events. Furthermore, the three unstable pitch events that were observed during the 30-minute flight occurred after the aircraft was out of the shadows of clouds where the solar power was essentially constant.

1.3.2 Aeroelastic / FCS Factors (TRUE)
It was determined that wing twist and lift loss due to elevator deflection, consequences of the aeroelastic characteristics of the aircraft and the FCS, contributed to causing the persistent high wing dihedral. It was also determined that the FCS's response and the wing's aeroelastic response to perturbations were found to not only increase wing dihedral, but also to sustain the dihedral. These two effects oppose the wing's return to lower, wing tip deflections.

The FCS reacts to the wing dihedral increase by deflecting the elevators trailing edge up. The elevators on the outer 40 feet of each wing are fixed. For any control law command requiring an elevator trailing edge up response, more lift is lost in the center of the aircraft than at the wing tips.

Also through video interpretation, separated airflow on the center wing section did not contribute to the persistent high dihedral. The spanwise redistribution of lift from the center of the aircraft to the outboard panels caused by the loss of lift in the center panel is reinforced by the wing tips twisting leading edge up as dihedral increases. The primary cause of dihedral-induced twisting of the wing tip leading edges is the relatively forward center of gravity and the elastic axis of the wing being above the chord line of the wing. This interrelationship forces the wing to twist leading edge up as the tips bend up. Figure 9.4 provides time histories of the wing dihedral and the elevator position.

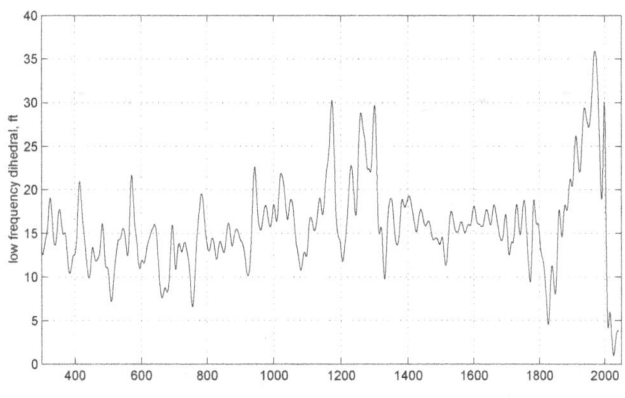

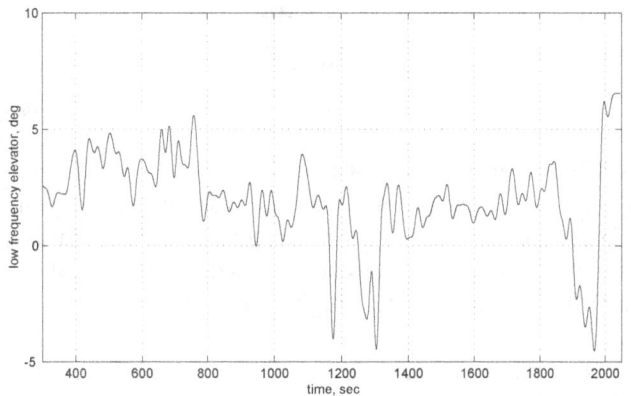

Figure 9.4 – Time History of Wing Dihedral and Elevator Position

The quantification of the primary effects of dihedral on elevator position and twist is best illustrated by analyzing the spanwise lift distribution. For a given airspeed and motor power setting, the predicted spanwise distribution of lift is provided in Figure 9.5. The spanwise redistribution of lift from the center of the aircraft to more outboard panels is the primary cause of the unexpectedly slow return to lower dihedral after any perturbation.

65

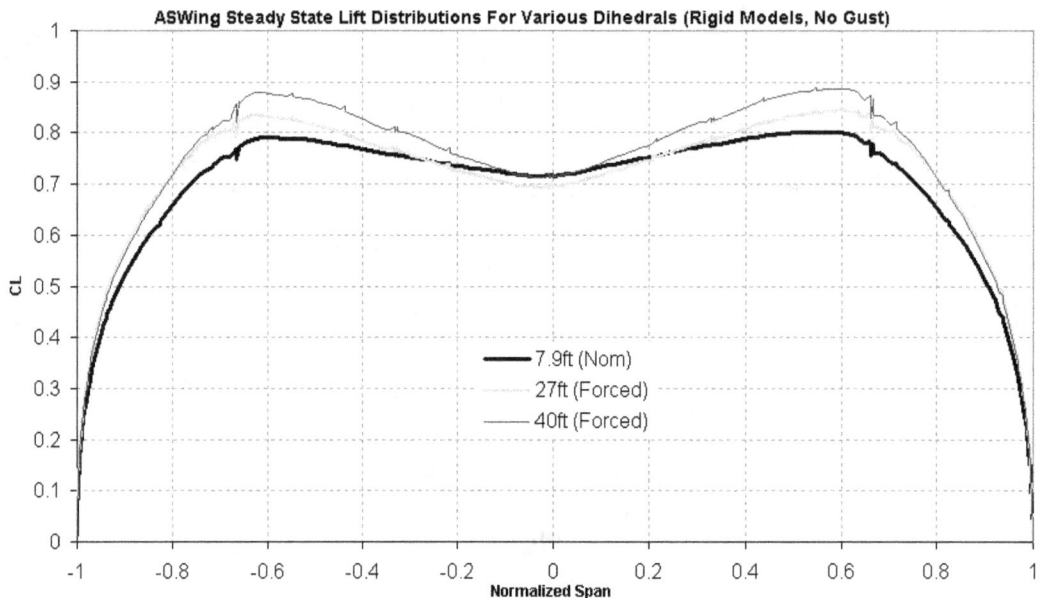

Figure 9.5 – Spanwise Lift Distribution for Various Wing Dihedral

1.3.3 Propulsion Factors (FALSE)

Four potential ways for the propulsion system to cause persistent high dihedral were explored. These included: propeller pitch, torque limiting, thrust response, and erroneous pitch rate sensing for thrust. It was determined that none of these significantly affected the persistent high dihedral.

- Propeller Pitch (FALSE)

"Propeller pitch was reduced by 2.5 degrees, as summarized in Document #13 found in Volume III, Appendix E, making it –8 degrees from initial design. This change allowed the motors to absorb more power at lower altitudes, which was required to compensate for reducing throttle commands to the tip motors by 50%. This change also improved climb performance by allowing use of more solar array power before encountering torque-limiting. Post-flight analysis, reported in Document #26 found in Volume III, Appendix E, indicates that this change in conjunction with the other between-flight changes had the desired effect on dihedral and did not contribute significantly to the pitch instability". (Excerpt taken from Document #25 found in Volume III, Appendix E, page 38)

- Torque Limiting (FALSE)

Motor torque limiting was analyzed as a potential contributing cause. The results of that analysis are summarized in Document #10 found in Volume III, Appendix E. These analyses indicated that the contribution of torque limiting to either high dihedral or pitch oscillations was minor.

- Thrust Response (FALSE)

"Throttle command to the two outboard motors was limited to 50% of center motors, as summarized in Document #13 found in Volume III, Appendix E. The tip motors received no drag command from the throttle. Turn command gains to the tip motors remained the same. This change was intended to reduce the effect of power changes on dihedral. Post-flight analysis of average dihedral at different power levels during the flight indicates that this change had the desired effect." (Excerpt taken from Document #25 found in Volume III, Appendix E, page 38)

- Erroneous Pitch Rate Sensing for Thrust (FALSE)
"Pitch and yaw rate sensors are located in Pod #2, located 41 feet to the left of the aircraft centerline, a detail that is not modeled in the ASWing code. At high dihedral, the sensors are rotated a few degrees in roll, which means that actual pitch rates are also sensed by the yaw sensors. The apparent measured yaw rate is a small fraction of the pitch rate, but it is large relative to "normal" yaw rates. That means that the aircraft reacts with a large differential throttle command. During Flight H03-1, the first flight, a mild pitch oscillation was apparent during lateral control sweeps. Thus it is possible that the very large steering commands evident during the final pitch oscillation could have fed back into the pitch rate feedback loops. This theory is not considered a primary cause of the instability, since it can be explained by previous analysis." (Excerpt taken from Document #25 found in Volume III, Appendix E, page 38).

1.3.4 Three Point Mass Dynamics (TRUE)
It was determined that adding three point masses to the HP01 configuration contributed to causing the persistent high dihedral. Although the Helios aircraft was conceived as a very simple aircraft design for high altitude solar flight, the structural flexibility and the large masses associated with the fuel cell system introduced substantial complexity into the aircraft's flight dynamics. The subset of complexities, which are relevant to understanding the mishap, is the relationship between wing dihedral, the addition of 520 lbs to the aircraft's centerline, the addition of 165 lbs near each wing tip, and the FCS. The large center mass required that the 165 lb hydrogen fuel tanks be placed outboard on the wing to prevent excessive wing dihedral. For the HP03 configuration at normal airspeeds and no turbulence, the wing dihedral varied on an average from about 11 feet to about 17 feet tip deflection during the first flight on June 7, 2003.

It is concluded that the persistent high dihedral caused the pitch instability. The pitch oscillation shares many characteristics with a traditional neutrally damped phugoid response of conventional fixed wing aircraft. For the Helios, the period of the neutrally damped pitch oscillation was about 8 to 9 seconds and the pitch rate was about 5 degrees/second. The stability of the oscillation was predicted and observed to be proportional to the wing dihedral of the aircraft. For wing dihedrals of about 30 feet wing tip deflection, the pitch oscillation became dynamically unstable. The predictions of the ASWing and Matlab codes, and flight estimates are shown in Figure 9.6.

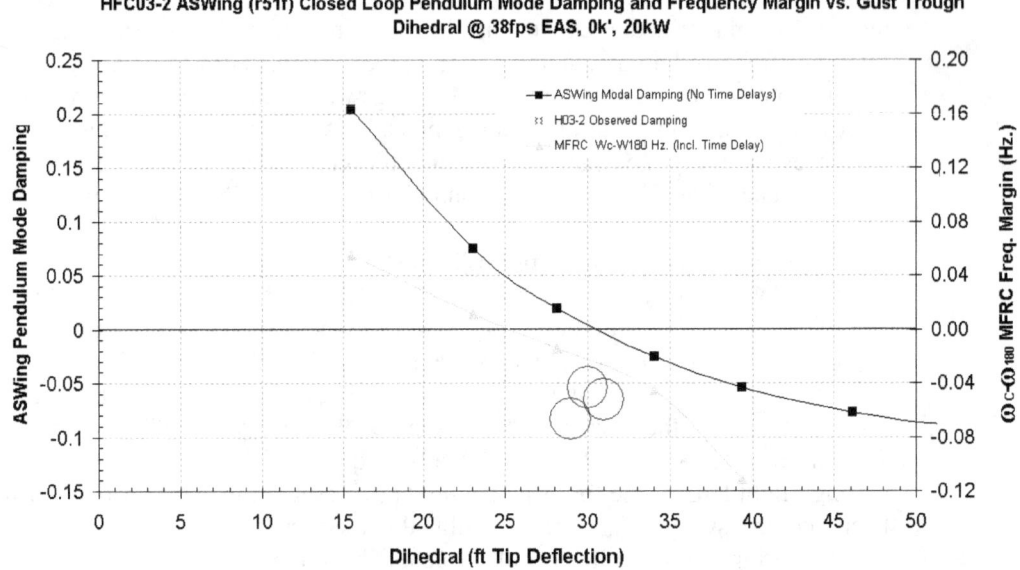

Figure 9.6 - ASWing Predicted Phugoid Mode Damping for Steady State Flight at 38 ft/sec

The Matlab predictions, which included time delays determined during the flight test, agreed better with the flight observations. Although both models predicted more stability than observed in flight, they clearly capture the instability. The difference between predicted and observed neutral damping is not considered significant because the differences are about the same as the uncertainty of estimating wing tip deflection from strain gauge measurements. Clearly, the light to neutrally damped oscillations observed in an event 5 to 6 minutes before the crash corroborates these predictions. Figure 9.7 and Figure 9.8 graphically captures the damping estimates represented by two of the blue circles (first encounter or event involving the unstable mode and the mishap event) in Figure 9.6.

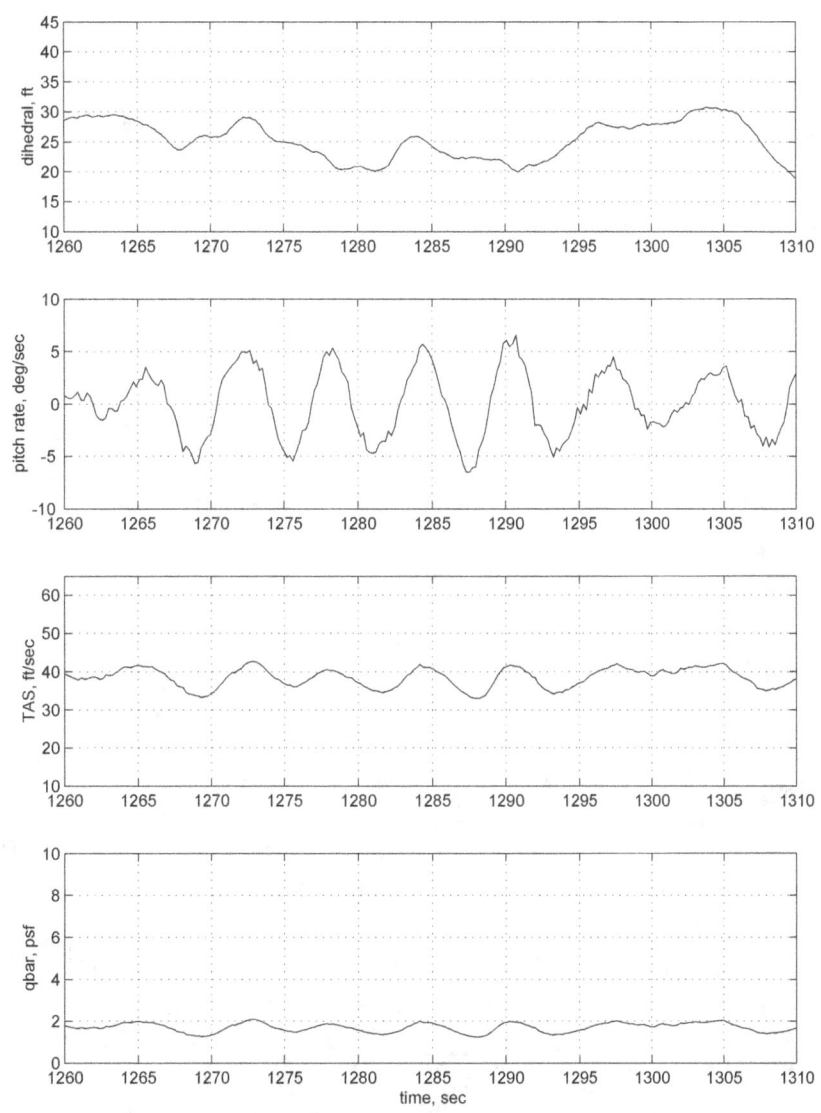

Figure 9.7 - Dihedral, Pitch, Vehicle Airspeed, and Dynamic Pressure for Event 1

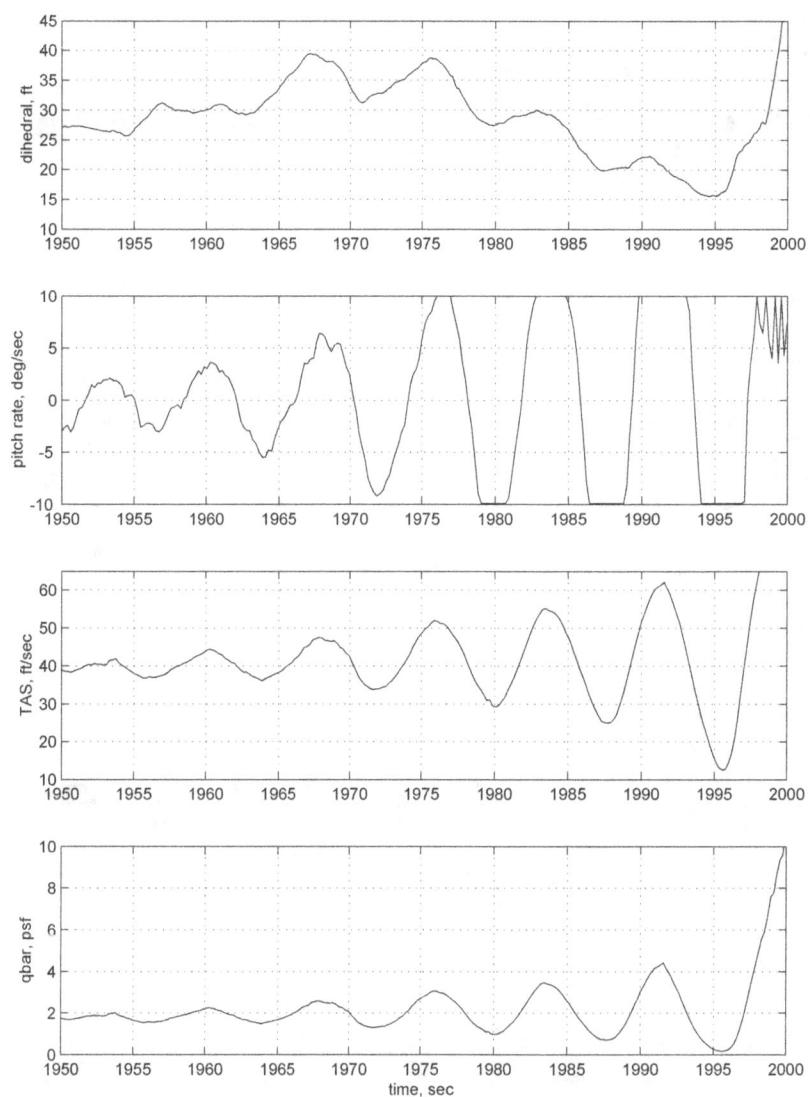

Figure 9.8 - Dihedral, Pitch, Vehicle Airspeed, and Dynamic Pressure for Mishap Event

The underlying physics of the phugoid mode and its potential instability is thought to be a consequence of two primary factors. First, large longitudinal static stability that arises from a high aerodynamic pressure relative to center of mass as the wing tips bend up. As the dihedral increases, the center of aerodynamic force moves further above the center of gravity. The growth in static margin with increasing dihedral is shown in Figure 9.9. This increasing static stability is believed to increase phugoid instability.

70

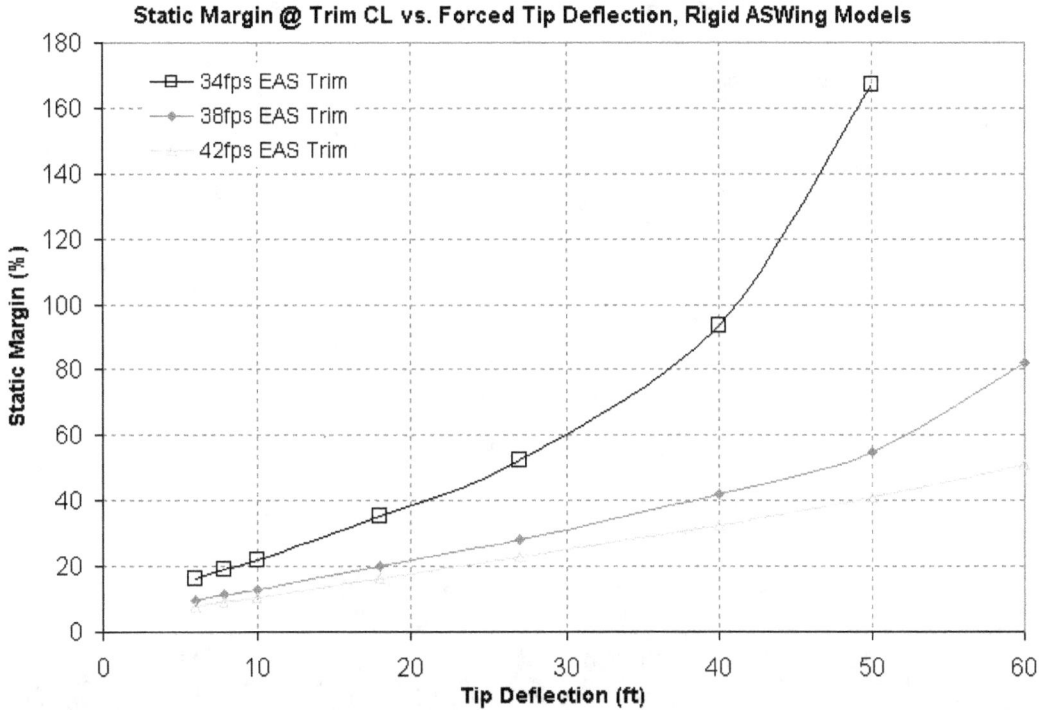

Figure 9.9 – Static Margin Versus Aircraft Stability with Increasing Airspeed

The second factor is the pitching inertia of the aircraft which increases dramatically while the restoring aerodynamic forces grow much slower. Although the dynamics of the HP03-2 pitch oscillation while in forward lifting flight are quite complex, a dominant force is the pitching inertia, which grew by a factor of five over the dihedral range seen on the HP03-2 flight. Figure 9.10 graphically relates the variation of pitch inertia with dihedral.

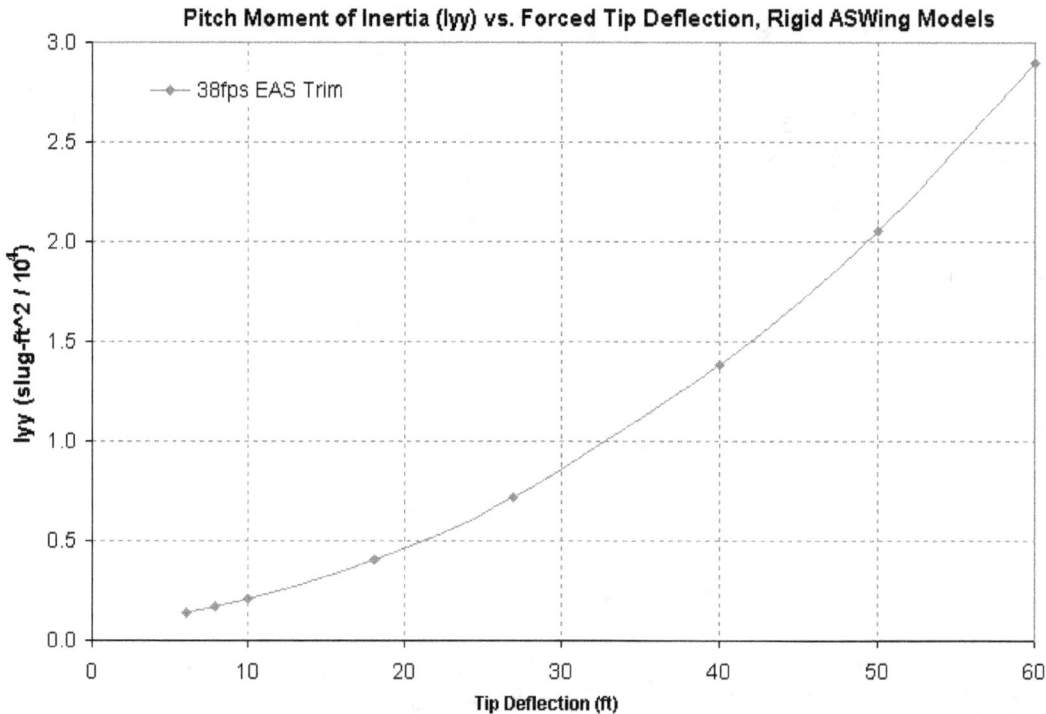

Figure 9.10 - Aircraft Pitch Inertia Versus Wing Dihedral at 38 ft/sec

At the large dihedrals, what may resemble a conventional neutrally damped, phugoid of a conventional aircraft is probably more accurately characterized as a dynamically unstable pendulum mode. The mode increases pitch attitude excursions, and increases airspeed with each nose down cycle despite apparent reductions in dihedral below the approximate 30-foot stability threshold. Those airspeed excursions quickly exceeded the maximum dynamic pressure design limit and resulted in the solar cell skin failures and subsequent loss of aircraft. The large wing tip deflections and more important, the persistence of dihedral was the physical cause of aircraft loss. Understanding the unanticipated large and persistent dihedral was a primary focus of post flight analytical analysis. The rapidly increasing airspeed excursions with each nose down pitch are obvious in Figure 9.8, the mishap event time history.

2.0 Decision to Fly Inappropriate (TRUE)

The MIB reviewed: 1) the rationale considered and the processes followed by NASA and AV in deciding that a single vehicle would be used for demonstrating the two ERAST missions (100,000 foot high-altitude mission using solar cells and the 50,000 foot altitude, 96-hour long-duration mission using a hydrogen-air fuel cell system and solar cells); and 2) the "day-of" procedures and decisions that resulted in flying the HP03-2 on 26 June 2003.

The Memorandum of Agreement (MOA) between AV and DFRC apportioned flight safety responsibilities on this program to AV and the mission success to DFRC. No specific NASA requirements were specified for completion of AV's airworthiness and flight safety responsibilities. Despite this broad direction, DFRC including senior management, clearly involved themselves in the review of critical information that would have collectively met the specific requirements of an Airworthiness and Flight Safety Review Board (AFSRB) as outlined in DFRC Handbook X-001.

Typically when DFRC has airworthiness and flight safety responsibility for a high-risk technology program, it typically invokes an Independent Review Team (IRT) to assess and present airworthiness to the DFRC AFSRB. Though the project team had followed a sound process with appropriate safety criteria for the "day of" decision to fly, the programmatic and technical decisions leading to the approval to fly this specific configuration of Helios suffered from incorrect assessment of risk as the result of inaccurate information provided by the analysis methods used, as well as, schedule pressures and fiscal constraints resulting from budgetary contractions, constrained test windows, and a terminating program. Though the pressures and constraints were not considered unusual for typical research projects at NASA, it did have some unquantifiable influence in the decision process.

2.1 Decision to Fly Configuration Inappropriate (TRUE)

Two key and sequential decisions (described in the next two paragraphs) were critical in leading to the disturbance-sensitive flight vehicle involved in the mishap. These decisions, which directly contribute to the root causes for this mishap, were spread over several years under considerable, but not unusual programmatic pressures. Basically, these decisions resulted in a spanloader configuration (properties nearly constant from wing tip to wing tip) being changed to a highly flexible aircraft with three quite heavy masses with large pitch inertias distributed along the aircraft without any increase in wing stiffness.

Decision to Fly a Single Aircraft: The original ERAST plan was to build two independent airframes: one optimized to satisfy the high-altitude mission using solar cells and one for demonstrating the long-endurance mission using a RFCS and solar cells. However, in early 1999 under the constraint of a reduced budget, NASA and AV agreed that using a single airframe to demonstrate both missions was necessary. The Helios Prototype became the single airframe for demonstrating the 100,000-feet milestone (HP01) in 2001 and the 50,000-foot, 96-hr fuel cell milestone (HP03) in 2003. The HP01 design would be a derivative of the Centurion aircraft and would be optimized around the high-altitude requirements. The long-endurance objective would be accomplished using a modified version of the HP01 configuration with the "then-planned" two RFCS pods installed at about 1/3 the distance from the aircraft center to the wing tip.

Decision to Eliminate the RFCS: The second decision made in 2001 and driven by available technology and remaining schedule, was to demonstrate the long-endurance goal using a primary fuel cell system rather than a regenerative full cell system.

The RFCS, which consisted of 2 fuel cell pods, was to be used for the long-endurance flight demonstration. As noted above, the original plan was to build a second airframe optimized for the loading of the 2 fuel cell pods. When the Centurion was modified to the six-panel configuration, the loads due to the fuel cell pods were considered in sizing the new center panels. Analysis indicated that the existing wing panels outboard of the fuel cell pods, would be strong enough if the designed-in-15% margin was used as growth. During October-November 2001, NASA and AV independent technical reviews were held to assess the progress in the RFCS development and to assess the actual Helios flight performance from the 2001 high-altitude flights. Based on this review and concerns about developing a flight-worthy RFCS in the available time, AV approached DFRC management with a proposal to change from a RFCS to a consumable PFCS for the Helios 2003 mission. In December 2001, NASA and AV decided to switch to the PFCS and began the PFCS development and modifications to the Helios aircraft for the 2003 mission. The PFCS required the use of three large masses (165 lb hydrogen fuel tanks mounted on the wing tip panels and the 520 lb fuel cell system mounted on the aircraft's centerline). These three masses weighed nearly 50 percent of the weight of the HP01 flown in 2001.

This second decision was the most critical to the aircraft's stability and control characteristics and greatly increased the likelihood of an unrecoverable state developing.

2.1.1 Risk Review Process Failed (FALSE)
A PDR and a CDR for the HP03 vehicle with the PFCS were accomplished in February 2002 and August 2002, respectively. Also during November 2002 through January 2003, technical reviews that involved independent discipline experts were completed to assess the HP03 changes, and to review the structural loads, stability and control, and aeroelastic modeling activities and corresponding predictions. Based on the modeling tools available and the predictions on aircraft behavior the decisions to fly the HP03 in the summer of 2003 were reasonable. DFRC "Technical Reviews" as outlined in DFRC Handbook X-001 were used as primary vehicles for NASA approval of configuration changes from Flight HP03-1 to the mishap flight.

2.1.1.1 Risk Review Process Does Not Exist (FALSE)
The MOA between AV and DFRC apportioned flight safety responsibilities on this program to AV and the mission success to DFRC. A risk process existed and was properly employed in determining the suitability of the HP03 configuration relative to these responsibilities. AV used a series of design reviews and a hazard management system to manage risk associated with the technical and operational aspects of the program. No specific NASA requirements were outlined in the MOA for completion of AV's airworthiness and flight safety responsibilities. Despite this broad direction, DFRC including senior management, clearly involved themselves in the review of critical information that would have collectively met (if not exceeded) the specific requirements of a formal AFSRB as outlined in DFRC Handbook X-001. DFRC Technical Review's in lieu of the DFRC AFSRB were employed as an oversight measure for DFRC's mission success role. The team had a well-documented and conscientious hazard management process.

2.1.1.2 Risk Review Process Inadequate (FALSE)
In general, the program suffered budgetary constraints that led to re-scoping of the technical options in the program. This re-scoping merged several counter-posing technical requirements that forced the vehicle design into areas where the

technical margins were narrow. Thus the behavior of the vehicle became extremely nonlinear and complex, and its dynamic response to disturbances difficult to predict. These earlier reviews/decision points lacked any real technical content in the area of stability and control (S&C), though it was acknowledged as a key factor in the projects successful outcome. The amount of external or independent review was marginal at best. The project would have been met with greater scrutiny, perhaps through an internal DFRC IRT, if DFRC had the flight safety responsibility for the project. The team and the Headquarters Enterprise did recognize the need for a thorough technical scrub of these preliminary decisions, launching into a set of technical reviews in the December 2001 and 2002 time frame.

During the last few years prior to the mishap, several factors led to a priority being established for demonstrating the commercial viability of the long-endurance vehicle. Management focus was shifted more to the Energy Storage System and away from the S&C of the vehicle, though clearly S&C was not neglected. External and/or independent review in these final years was largely nonexistent and consisted mainly of DFRC's AFSRB. As the program approached the flight test window during the summer of 2003, it was apparent that a major milestone had to be accomplished by September 2003 without the possibility of schedule or budget relief.

Despite these apparent pressures, the team of AV and the DFRC project office conducted a series of reviews (see Table 9.2) with marginal independent technical oversight over several years. AV accepted these reviews as beneficial and embraced the reviews with a spirit of cooperation and enthusiasm. DFRC's senior management was engaged in the reviews and programmatic decision-making processes as a result of the DFRC's program manager's effort to satisfy DFRC's responsibilities and keep management informed. AV's technical review process was the basis for all major technical and programmatic decisions. Each decision point in the configuration's evolutionary path was assessed by these reviews with all "apparent factors" on the table.

It was recognized by the team that this design clearly embarked into the "research" area of these interrelated disciplines, especially at the higher dihedral conditions. As such the possibility of a short technical review uncovering an unknown sensitivity to disturbances or vehicle responses that lead to sustained high dihedral are low at best. Sustained technical involvement or in-depth and independent technical analyses are meaningful alternatives that might significantly reduce the risk in this area.

Table 9.2 – Internal and External Reviews

Review Title	Date	Purpose/Content	DFRC Participation	Independent Review
Project Redirection	Feb 99	Scale project to single vehicle concept/No S&C	Project Management	None
Helios Conceptual Design Review	May 99	Vehicle prototype with RFCS design review/ No S&C	DFRC Project Management and Engineering; S&MA; GRC (fuel cell)	None
Preliminary Design Review	Sep 99	Vehicle prototype with RFCS design review	DFRC Project Management; Range Safety; GRC (fuel cell)	LaRC S&C (1)
ESS Independent Management Review	Oct 99	Review ESS Programmatics; No S&C	Project Management	SRS and SAIC
Independent Technical Review	Sep 00	Review Helios Technical Concept/No S&C	Project Management; Flight Operations	Naval Research Lab, MSC, Aerospace Corp, HQ NASA
Loads & S&C Review	Dec 01	Vehicle loads and stability & control review	DFRC Project Management and Engineering; DFRC Non-project engineering (4)	LaRC S&C (1)
Project Redirection	Dec 01	Shift from RFSC to PFCS design	DFRC Management	None
Preliminary S&C, Aeroelasticity and Dihedral Analysis Review	Dec 02	S&C, Aeroelasticity and Dihedral Analysis Review	DFRC Project Management and Engineering; DFRC Non-project engineering (4)	DFRC consultant (1); LaRC (2)
Final review of S&C, Aeroelasticity and Dihedral Analysis	Jan 03	Review final design analysis	DFRC Project Management and Engineering; System Safety	None
Mission Success Briefing	Feb 03	Review design and operational approach for Summer 03 flight test	DFRC AFSRB membership	None

Review Title	Date	Purpose/Content	DFRC Participation	Independent Review
Deployment Readiness Review Part I	Mar 03	Assess Hydrogen Tank and support Equipment design and operational readiness to deploy to Hawaii for Summer '03 flight test	DFRC Project Management and Engineering; System Safety; DFRC Flight Operations (1)	None
Deployment Readiness Review Part II	Apr 03	Assess team and aircraft and Fuel Cell design and operational readiness to deploy to Hawaii for Summer '03 flight test	DFRC Project Management Engineering; System Safety; DFRC Flight Operations (3); S&MA Office (1)	None
Tech Brief for Flight HP03-01	30 May 03	Assess project's technical, operational and safety readiness for flight	DFRC AFSRB members; DFRC Project Management Engineering; System Safety	None
H03-01 Crew Brief	5 Jun 03	Coordinate and assess final team and vehicle readiness for flight	Team and PMRF personnel	None
Tech Brief for Flight HP03-2	24 Jun 03	Assess project's technical, operational and safety readiness for flight of last configuration changes	DFRC AFSRB members; DFRC Project Management Engineering; System Safety	None
HP03-02 Crew Brief	24 Jun 03	Coordinate and assess final team and vehicle readiness for flight	Team and PMRF personnel	None

During the period of January 2003 to April 2003, the design, development, and fabrication of the PFCS pod was completed, and the HP03 aircraft was modified. All PFCS testing and aircraft subsystem functional testing was completed. During February 2003, a MSR was held with NASA and AV to review the final aircraft configuration changes, flight operations approach, risk management process, systems safety, and program controls. On 12 March 2003 the first DRR was conducted to assess the system design modifications, qualification testing, airworthiness, and operational readiness for the hydrogen tanks and ground support equipment, and on 8 April 2003 a second DRR was conducted. Based on these reviews, it is believed that the risk process used by NASA and AV was adequate. There was no evidence that either AV or the government ignored exposed risks. It is the opinion of the Board that AV and the DFRC project office were willing to slip the schedule or miss a milestone if it was necessary to address safety-related issues. Senior government management was engaged in the decision process at both DFRC and the associated NASA Enterprise. The team provided appropriate detail of the information as reviews progressed. There were no indications that known risks of the program were inappropriately miscommunicated.

2.1.2 Review Content Was Inadequate (TRUE)
Three specific factors contributing to insufficient technical design and analysis were explored. These included: math model fidelity, design and analysis tools adequacy, and model validation. The Board concluded that since the pitch instability was predicted when the aircraft was at high dihedral, the fidelity of the math models was adequate. However, the tools and the solution techniques were deemed to be inadequate for predicting the vehicle's sensitivity to disturbances and lack of robustness to return to a low dihedral condition. Additionally, validating math models and predictions using ground and flight test data was found to be inadequate.

For this investigation, design and analysis tools are defined as the codes and solution techniques used by the engineers and scientists. Examples of tools include: NASTRAN, ASWing, Matlab, and Mathematica. Math models are defined as data sets required by a code to perform some analysis or design function.

2.1.2.1. Math Models Have Inadequate Fidelity (FALSE)
Five math models that were used to design and analyze the vehicle were assessed to determine if they had adequate fidelity. It was determined that the environmental, dynamic, aerodynamic, propulsion, and structural models did have adequate fidelity since they did predict the instability at high dihedral. Although the <u>persistent high dihedral</u> and the <u>highly divergent nature</u> of the unstable pitch mode were not predicted prior to flight using the design and analysis tools available at that time, a persistent high dihedral was predicted when these models were used with time domain analysis tools enhanced by AV during the mishap investigation. Accurately predicting damping away from the neutrally stable condition (damping equals zero) has always been difficult using linear frequency domain techniques. The use of time domain techniques promises more accurate predictions of damping.

2.1.2.2 Design and Analysis Tools Inadequate (TRUE)
"The continuous turbulence model analysis (NASTRAN and Mathematica models) that was performed prior to the 2003 flight tests indicated that the centerline bending moment sensitivity for this form of gust was approximately 15% higher than it had been for the HP100k aircraft. This linear analysis actually matched the HFC03-2 ASWing results fairly closely for large gusts, but it did not capture the non-linear behavior seen for small gust amplitudes" (excerpt taken from Document #25 found in Volume III, Appendix E, page 23). In addition, the complexity and interactions between aeroelastic and stability modes made it difficult to apply time domain stability and control analysis to the vehicle. This difficulty forced program engineering to rely primarily on static stability and

frequency domain analyses, which did not show the time dependent interrelationship between wing twist and FCS reactions and dihedral.

2.1.2.3 Model Validation Incomplete (TRUE)

Uncertainty in the predictions was recognized, but ground and flight test validation of the predictions and models were not pursued due to programmatic constraints. As an example, a recommendation was offered to the team by an independent technical advisor to perform a buildup in weight to assess the rate of stability and control margin degradation and vehicle robustness to extreme dihedral excursions. This represented an excellent mitigation to the vehicle's susceptibility to and probability of encountering some unknown, undesirable phenomenon. But because of fiscal and schedule constraints, the team had to declined the extra effort to validate the design, math models, and vehicle performance.

At the present time there are no established test techniques or procedures to conduct ground vibration testing of such large elastic structures that will accurately calibrate analytical tools.

2.1.3 Risk Categorized Improperly (TRUE)

As part of the risk management process, the hazard management process did identify potential hazards. The team's hazard management process was consistent with NASA standards and produced a set of hazards for ground, flight, and range operations. The hazard reports reflected a well documented and thorough process based on team deliberations and included significant lists of mitigations. There was signed documentation certifying compliance with the proposed mitigations, however the document did not provide an audit trail that would substantiate such claims. A review of AV's thorough documentation of analyses and decisions verified intended mitigation completion.

Two flight hazards were identified and considered relevant to the longitudinal instability experienced on the mishap flight (see Appendix G, Volume III). These were "Poor Stability and Control" and "Structural Failure/Damage". Though several causes were listed, the first hazard report failed to identify the specific mechanism leading to an unstable and quickly diverging phugoid response, as well as, the interactive effects of many factors that might contribute to realization of such an outcome. The probability of this hazard was assessed in its mitigated state as "remote". The second hazard report recognized the possibility that weather could directly exceed design limitations and was assessed as more likely than an instability and divergent control response.

Absent from all hazard reports was any evidence of the probability of occurrence that supported both the unmitigated and mitigated probability of occurrence for the hazards. Typical to many research programs, engineering judgments are the primary and sole source that affixes the risk category according to DFRC's risk matrix. The team and management reviews did highlight a significant change in the vehicle's weight and balance that would affect both dihedral and static and dynamic stability. However, the additional potential risk was not reflected in any change in the risk occurrence categorization despite occurrences of some longitudinal divergent oscillations that persisted for less than a couple of cycles before dampening out on previous flights under more benign conditions and configurations. There was no indication in DFRC's management documentation of their recognition of this increased risk above the briefed levels.

Overall however, DFRC's "Technical" and "Operational Readiness" Reviews, which represented the most proximate reviews to the mishap flight involving DFRC senior management, characterized the developing technology as "complex and inherently risky", and the risk to mission success as "moderate to high".

2.2 Decision to Fly Mishap Sortie Inappropriate (FALSE)
The procedures developed and used by AV in preparing for the HP03-2 flight are quite extensive and were followed very closely. Various formal meetings between NASA project and AV personnel are defined in Table 6.1 and summarized in Document #19 found in Volume III, Appendix E. The procedures are documented in an AV report entitled "Helios Fuel Cell Flight (H03-2)" dated 26 June 2003. All aircraft, system, and equipment checks were nominal. Based on this information, the decision to fly the HP03-2 flight was appropriate.

2.2.1 Risk Review Process Failed (FALSE)
Many meetings took place between May 2003 and 26 June 2003 to plan for, review, and conduct three flight experiments, namely the straight-line flight conducted on 15 May 2003, the first long-endurance flight conducted on 7 June 2003, and the second long-endurance flight (flight of mishap) conducted on 26 June 2003.

2.2.1.1 No Risk Management Exists (FALSE)
The project team used a set of launch and mission success criteria to successfully manage the day-of hazards. These included not only go/no-go operational limitations and real-time data requirements in the control room, but structured preflight and post-flight crew briefings. All these techniques were reinforced by set of disciplined checklists for each crew position and a detailed mission script.

2.2.1.2 Risk Management Process Inadequate (FALSE)
The straight-line flight test was completed to verify the proper wing dihedral distribution of aircraft. A Technical Briefing between NASA and AV personnel was conducted on 30 May 2003 to review the ground and straight-line flight test results, flight plan, configuration changes, operational procedures, and system safety, and to close out any remaining items prior to the first June flight.

On 5 June 2003 the HP03-1 crew briefing was held at which time test objectives, final aircraft status, test timeline, safety issues, roles and responsibilities, flight plan, and the weather forecast were reviewed. The first high-altitude flight of the HP03 was conducted on 7 June to validate the handling and aeroelastic stability of the aircraft with its fuel cell system and gaseous hydrogen storage tanks installed, and to obtain data on the performance of the fuel cell system in the stratosphere.

On 24 June 2003 a Technical Briefing between NASA and AV personnel was conducted to review the HP03-1 test results, flight plan, configuration changes, operational procedures, and system safety, and to close out of any remaining items prior to the flight scheduled for 26 June. Based on an assessment of the risk, all NASA, AV, and PMRF personnel supporting the flight test indicated a "Go" for flight commencing on 26 June.

The crew conducted and documented a day-prior briefing for the flight. Final parameters relative to weather, data requirements, and crew station and vehicle functionality were assessed prior to takeoff. No evidence was found to indicate that the team violated any limitations or bypassed any procedural steps.

2.2.2 Review Content Inadequate (FALSE)
The content of all reviews and briefings during the period May through June 2003 was considered adequate.

3.0 Pilot Fails to Stabilize (TRUE)
Two potential causes of the pilot not preventing the mishap were explored. These included an assessment of the pilot's ability to control the instability, and an assessment of the pilot's use of procedures available to control the instability.

3.1 Instability Beyond Capability (TRUE)
Failure of the pilot to stabilize the vehicle was due to design and test philosophies established for this very unique program and vehicle. The pilot's control panel was designed to provide only standard "autopilot type" mode and navigation inputs; it was not designed to provide for direct pilot-in-the-loop control of attitude nor was it designed to provide the pilot capability to recognize an impending departure from controlled flight or to stabilize the aircraft. Review of the wing tip video tapes of the mishap combined with analysis of the pilot's actions has led the MIB to conclude that any pilot actions taken after the aircraft entered into the divergent pitch oscillations at high dihedral could not have prevented the final dive and eventual secondary structural failure due to excessive airspeed.

3.2 Pilot Failed to Use Capability (TRUE)
The stability and control philosophy for the solar-aircraft is that the vehicle aerodynamics and FCS will keep the aircraft in stable flight without pilot input. The pilot should never have to maintain attitude or stabilize the aircraft in pitch. However, emergency procedures were developed for the situations in which the aircraft stability was not satisfactory. The most time-critical emergency procedures are called "non-deferred" procedures and must be handled immediately.

3.2.1 Failed to Use Proper Emergency Procedure (TRUE)
The intent of non-deferred emergency procedures (Appendix F, Document F.3) is to think through all identified high-risk scenarios and argue out a best course of action for

each situation. When the team of pilots and engineers agrees on the best sequence of actions, the procedure is written, practiced, and committed to memory. In the event of a time-critical emergency, the pilot alone makes the decision to execute a non-deferred emergency procedure and immediately performs the steps.

3.2.1.1 Pilot Chose Incorrect Checklist (TRUE)
The correct non-deferred emergency procedure for the mishap situation was the Long Period Pitch Oscillation procedure (see Appendix F, Document F.3). The pilot remarked at 10:36:17 that he was "in a phugoid of sorts, a big one". This was an indication that he had initially assessed and identified the emergency correctly. However, at 10:36:22 the SP stated that the extreme pitch attitudes and sharp stall break led him to conclude that the emergency was actually a nose-up pitch hard over. He asked other crewmembers for advice and after receiving none, initiated the Pitch Hard-Over non-deferred emergency procedure (see Appendix F, Document F.3). This was not the correct procedure for the actual situation. The first step defined in the Pitch Hard-Over non-deferred emergency procedure required the pilot to turn the airspeed hold off. Selecting the wrong procedure by the pilot was result of not correctly identifying the impending aircraft response when the airspeed hold command is turned off.

3.2.1.2 No Assigned Checklist (FALSE)
The Long Period Pitch Oscillation was the correct "non-deferred" procedure that most closely matches the mishap condition. This procedure was intended to be used to control the aircraft response during a slowly divergent pitch oscillation in which airspeed would be the primary controller and the time-to-recognize and respond would be generous.

3.2.2 Failed to Respond in a Timely Manner (TRUE)
The crew could not see pitch oscillations in the wing camera view due to the high dihedral, which moved the horizon outside the viewing area, nor did the S&CE recognize the pitch oscillation growth amid the turbulence. The Dynamics Engineer (DE) also did not report the strain gauge exceedances to the SP until asked about them. The camera view (spanwise from the right wingtip) kept the crew focused on the high dihedral, and slowed any awareness of the developing oscillation. As a result, the crew's situational awareness and reaction time was well behind the actual aircraft flight behavior.

Numerous cautionary factors were known before the flight. If assessed together they should have peaked the team into a much greater *situational awareness,* and sparked a very elevated sensitivity to unusual aircraft behavior and a tough discussion regarding what immediate actions should follow if any anomalies were observed. These cautionary factors include:

1) Math models predicted that wing dihedrals greater than about 30 feet would cause an unstable pitch oscillation to occur. These models also predicted that the HP03 configuration was more sensitive to disturbances than for the HP01 aircraft due to

increases in mass and airloads without an increase in structural stiffness. The new vehicle configuration merged several counter-posing technical requirements that forced the vehicle design into areas where the technical margins were narrow. Thus the behavior of the vehicle became extremely nonlinear and complex, and the aircraft's dynamic response difficult to predict.

2) The PFCS (installed on HP03) was largely untested in turbulence. No incremental flight-testing was performed and the only other flight with this major configuration change was flown on a day without normal trade wind conditions and with unusually weak flow.

3) Since HP03 flew at a somewhat higher equivalent air speed than previous solar-powered configurations without an increased rate of climb, the slope of the climbout trajectory was smaller than previous flights. This meant a longer flight trajectory over which the airplane was exposed to the greater turbulence at lower levels, and along which it was desirable to try to avoid the shear lines. It would also be exposed to more of the low level turbulence in Kauai's wake.

4) The flight team considered the weather on the flight day to be unusual based on more southerly direction of trade winds and their strength.

5) The meteorologist indicated during the Go-No Go meeting that the southern shear line was expected to be very close offshore. During the poll, the meteorologist was only a *"very marginal GO"* based on concerns about clouds, upper level turbulence, and shear lines during climbout.

6) Take off would be the latest since flights began in 1997; therefore, there would be more atmospheric uncertainty due to the more fully developed sea breeze and its interaction with the offshore trade wind flow, as well as more likelihood of turbulence due to thermals off the heated land.

7) At 9:38am, the meteorologist reported to the MP that Sodar one sigma vertical velocity values had increased from 0.4 m/s to as high as 0.8 m/s, the highest ever for take off. The flight constraint was 1.0 m/sec

8) At 9:36am, the tower reported that whitecaps were already visible offshore, indicative of the very close proximity of the southern shear line. Whitecaps are clearly visible in Figure 7.2.

During climbout the following wing strain measurements were observed by the DE:

    10:07:29am: 80%, equals highest ever seen
    10:20:22am: 85%
    10:22:59am: 103%
    10:33:34am: SP to DE: "Strains on the spars have been pretty negligible up to now"
              DE to SP: "Negative. --Seeing the biggest ever seen—"

83

10:35:53am: 115%, mishap Event 3 has begun
10:36:06am: 122%
10:36:32am: 137%

Considering the cautionary factors above, and the subsequent high strain readings, which began shortly after take off and became worse with no evidence of discussion by the flight crew, indicate the crew's lack of situational awareness.

3.2.2.1 Crew Fails to Recognize Instability (TRUE)

The crew failed to recognize the ever-increasing instability. This was exacerbated by the previous missions success and encounters that afforded long times for recognition and response. Performance predictions had not indicated such stability issues. For this reason, the crew was pre-conditioned by a benign first flight and subsequently surprised by the in-flight departure. Secondly, criteria and ill-defined crew roles and responsibilities concerning instability recognition compounded the problem. The lack of response to the pilot's initial inquiry concerning the pitch response was indicative of this problem. In addition, a photo helicopter witnessed the event and was manned by an aerospace engineer who was very familiar with the vehicle, but was not linked to a command and control frequency.

3.2.2.1.1 Pilot Failed to Recognize Instability (TRUE)

The pilot failed to recognize the instability because of inadequate training, task saturation, and inadequate pilot/aircraft displays and interface.

3.2.2.1.1.1 Training Was Inadequate for Recognition (TRUE)

Crew training was somewhat informal (see Appendix F, Document F.2 and F.4) and contributed to the failure to recognize the impending instability. While non-pilot crewmembers were present and involved in flight simulations, they served more as an advisory function on systems operation than as a participatory crewmember. The experimental nature of this vehicle contributed to an inadequate training model, which impacted recognition and performance. In addition, infrequent flight operations negatively impacted any training syllabus and learning continuity.

3.2.2.1.1.2 Pilot Was Task Saturated (TRUE)

Approximately 10 minutes prior to the event, the pilot was progressively becoming task-saturated with multiple demands on his attention. This included concurrent concerns with the flight hand-off procedure (matching the SP and MP GCS switch settings, see Appendix F, Document F.5), high dihedral, vehicle flight state, and excessive radio and interphone communications (see Appendix F, Document F.7).

84

3.2.2.1.1.3 Pilot Was Complacent (FALSE)

There were no indications that the pilot had become complacent about his duties or his knowledge of the aircraft systems (see Appendix F, Document F.1).

3.2.2.1.1.4 Pilot's Interface Inadequate (TRUE)

The pilot's primary display was designed to involve the pilot in selecting autopilot choices for navigation, the airspeed, the power setting, and FCS gains; it was not designed to allow direct control of aircraft attitude through elevator or power control.

Aircraft dynamic stability was determined to be a function of dihedral, which was typically controlled through the constraints of an airspeed envelope and the automatic actions of a pitch rate feedback stability augmentation system (SAS). As a result the pilot was not provided an attitude reference. However, the pilot was afforded visual cues through cameras that provided selectable views in orthogonal directions. Since part of the pilot's workload was to monitor the control system stability, the use of visual cues was crucial in early recognition of instabilities, however monitoring of dihedral and pitch stability were at odds to each other since each required a different camera view which was not possible simultaneously. Furthermore, the fidelity of the forward view suffered from poor horizon definition as a result of picture quality, haze, and the lack of terrain features for the test route. Typically, a trained pilot uses the horizon (or artificial horizon) for important cues for initiating recover from an aircraft departure.

3.2.2.1.2 Other Crewmembers Failed to Recognize Instability (TRUE)

The crew could not see pitch oscillations in the wing camera view due to the high dihedral, which moved the horizon outside the viewing area, nor did the S&CE recognize the pitch oscillation growth amid the turbulence.

3.2.2.2 Procedure too Long to Execute (FALSE)

In general, once the pilot recognizes an unacceptable situation requiring the use of a non-deferred emergency procedure, the steps required of the pilot are minimal because there are only 3 to 4 steps (see Appendix F, Document F.3) and they are memorized and can be accomplished in a matter of seconds (see 3.2.1). For this mishap, the wrong non-deferred emergency procedure was selected as described in 3.2.1.1, so an assessment of the correct non-deferred procedure being too long to execute could not be made.

3.2.3 Failed to Apply Proper Emergency Procedure Correctly (TRUE)

The proper emergency procedure was not selected by the pilot (see 3.2.1); therefore the proper emergency procedure could not have been applied correctly.

4.0 FCS Fails To Stabilize Instability (TRUE)

The FCS was designed to stabilize the unstable phugoid mode within its defined operational envelope. Outside of this envelope, the FCS was predicted to be ineffective in controlling the pitch instability. For this vehicle, wing dihedral was the key parameter defining the operational envelope. It was predicted that for a wing dihedral of less than 30 feet the unstable phugoid mode could be controlled by the FCS; above 30 feet the FCS would not be effective.

4.1 FCS Fails Outside Envelope (TRUE)

Above 30 feet of dihedral the pitch control law does not have sufficient authority to stabilize a pitch instability. At high dihedral, two branches of the pitch control law effectively cancel each other out leaving only an integrated pitch rate signal going to the elevator. The stability margins of this specific FCS implementation were analyzed and correctly identified for high dihedral conditions. Since the design of this control loop was inadequate to stabilize instabilities known to exist above 30 feet of dihedral, this condition defined the boundary of the aircraft's flight envelope.

4.2 FCS Fails Within Envelope (False)

Below 30 feet of dihedral the flight control system stabilized the vehicle. At these levels of dihedral the ASWing and Matlab codes predicted that the vehicle was closed-loop stable and flight test time histories show no pitch rate instability.

Section 10
Proximate Causes, Root Causes, Contributing Factors, Significant Observations, Findings, and Recommendations

This section of the report provides the results of the Board's investigation relative to the Proximate Causes, Root Causes, Contributing Factors, Significant Observations, and Findings. Definitions describing these factors are provided below.

Proximate Cause
The event that occurred, including any condition that existed immediately before the undesired outcome, that directly resulted in its occurrence and, if eliminated or modified, would have prevented the undesired outcome.

Root Causes
One of multiple factors (events, conditions, or organizational factors) that contributed to or created the proximate cause and subsequent undesired outcome and, if eliminated or modified, would have prevented the undesired outcome.

Contributing Factors
An event or condition that may have contributed to the occurrence of an undesired outcome but, if eliminated or modified, would not by itself have prevented the occurrence.

Significant Observations
A factor, event, or circumstance identified during the investigation that did not contribute to the mishap, but if left uncorrected has the potential to cause a mishap, injury, or increase the severity should a mishap occur.

Findings
A conclusion based on facts established during the investigation that was not identified as a root cause, contributing factor, or significant observation.

Section 7 provided a description of the events leading up to the instability and crash of the vehicle. The investigation determined that as the aircraft configuration evolved from a spanloader configuration to a configuration involving 3 large point masses, existing design and analysis tools failed to predict the vehicle's increased sensitivity to external disturbances. On 26 June 2003, the aircraft was perturbed by turbulence, morphed into an unexpected, persistent, high dihedral configuration that caused an unstable, highly divergent, pitch oscillation to occur from which vehicle recovery was not possible. During the pitch oscillation the aircraft experienced a high-speed dive that significantly exceeded the aircraft's design airspeed resulting in failure of secondary structure, and subsequently loss of lift.

The **Proximate Cause** for the loss of the HP03-2 was the high dynamic pressure reached by the aircraft during the last cycle of the unstable pitch oscillation leading to failure of the vehicle's secondary structure.

The Root Causes, Contributing Factors, Significant Observations, and Findings are presented in Tables 10.1 through 10.4. Recommendations are provided as bullets in each table, and then summarized in Table 10.5.

Table 10.1 - Root Causes and Recommendations

RC.1) Lack of adequate analysis methods led to an inaccurate risk assessment of the effects of configuration changes leading to an inappropriate decision to fly an aircraft configuration highly sensitive to disturbances.

- Develop more advanced, multidisciplinary (structures, aeroelastic, aerodynamics, atmospheric, materials, propulsion, controls, etc) *"time-domain"* analysis methods appropriate to highly flexible, "morphing" vehicles.

- Develop ground-test procedures and techniques appropriate to this class of vehicle to validate new analysis methods and predictions.

- For highly complex projects, improve the technical insight using the expertise available from all NASA Centers.

RC.2) Configuration changes to the aircraft, driven by programmatic and technological constraints, altered the aircraft from a spanloader to a highly point-loaded mass distribution on the same structure significantly reducing design robustness and margins of safety.

- Develop multidisciplinary (structures, aerodynamic, controls, etc) models, which can describe the nonlinear dynamic behavior of aircraft modifications or perform incremental flight-testing.

- Provide adequate resources to future programs for more incremental flight-testing when large configuration changes significantly deviate from the initial design concept.

Table 10.2 - Contributing Factors and Recommendations

CF.1) Flight control system was unable to control the pitch instability at high persistent wing dihedral.

- Consider implementing mitigations or hardware systems for returning a vehicle back into a safe flight envelope without catastrophic results when performing hazardous or envelope expansion testing but insure this does not increase overall risk of such testing.

CF.2) Hazard reports failed to provide any evidence that supported the probability of occurrence of the mishap associated hazards.

- Safety experts should explore the possibility of developing improved and believable criteria for more rigorous assessment of the probability of occurrence of identified hazards for one of a kind research aircraft.

CF.3) Pilot control module/interface lacked features that would afford the pilot the ability to recognize and mitigate an impending departure from controlled flight in a timely manner.

- Develop a human/vehicle system interface to better conduct research for this class of vehicle.

- Improve the pilot/crew displays to allow better recognition and situational awareness of slow developing hazardous events.

- Consider adding attitude indicator to improve pilot's situational awareness.

- Develop a method to measure wing dihedral in real-time with a visual display available to the test crew.

- Develop manual and/or automatic techniques to control wing dihedral in flight.

- Re-evaluate providing capability that allows pilot to better mitigate unusual flight occurrences.

CF.4) Pre-conditioning associated with previous successful yet infrequent flights of this class of vehicle, encounters with benign unstable phugoid responses, and informal crew training inhibited the team's ability to predict, identify, and react to the impending instability.

- Further refine the roles and responsibilities of the crewmembers to improve overall team response to unexpected and anticipated emergency conditions.

- Refine emergency recognition criteria to improve team emergency response.

- Perform simulations to develop recognition criteria that identify the vehicle's response prior to and during instabilities.

- Improve the fidelity of aircrew simulations to mitigate the risks associated with takeoff and landing.

CF.5) The review process was not structured to adequately identify the risks associated with this vehicle design especially as design margins were decreasing and complexity was increasing.

- Enhance the depth and independence of technical participation in the research areas of this class of vehicle.

CF.6) Pilot and crewmembers failed to recognize instability in a timely manner.

- Develop capability to perform simulations of the vehicle's response to disturbances.

- Improve training program and use simulations to enhance crew resource management during normal, emergency, and unstable flight conditions.

- Apply crew resource management techniques to enhance crew ability for identifying and responding to emergency and unstable flight conditions.

CF.7) Pilot was task-saturated, particularly during mishap event.

- Improve interfaces to alleviate pilot task-saturation.

- Re-evaluate pilot and test team responsibilities to optimize task management.

- Extend test team responsibilities to include more participatory tasks with provisions for providing advisory status of systems operation.

CF.8) Weather conditions as a function of altitude were not fully understood.

- Consider applying advanced atmospheric models that better predict conditions hazardous to this class of vehicle.

- If appropriate, re-evaluate the need to validate advanced atmospheric models using existing and, as necessary, special observations.

- Ensure that meteorological hazards in the area of operations are identified and understood.

Table 10.3 - Significant Observations and Recommendations

SO.1) Aerospace expert onboard photo helicopter not linked to a command and control frequency.

- If photo helicopters are available, consider providing capability for direct voice communication between the helicopter and the Helios pilot as long as this communication does not appreciably add to Helios pilot workload.

- Consider the chase plane concept of operations to improve overall test team management of unexpected and anticipated emergency or unstable conditions.

SO.2) Take-off and landing of this class of vehicle are high pilot/crew workload events with significant elevated risk.

Table 10.4 - Findings

F.1) Pilot selected the wrong emergency procedure to apply, however this did not cause the mishap.

F.2) Reliability of sensors and subsystems associated with FCS was a significant factor in design philosophy.

Table 10.5 – Recommendations

R.1) Develop more advanced, multidisciplinary (structures, aeroelastic, aerodynamics, atmospheric, materials, propulsion, controls, etc) *"time-domain"* analysis methods appropriate to highly flexible, "morphing" vehicles.

R.2) Develop ground-test procedures and techniques appropriate to this class of vehicle to validate new analysis methods and predictions.

R.3) For highly complex projects, improve the technical insight using the expertise available from all NASA Centers.

R.4) Develop multidisciplinary (structures, aerodynamic, controls, etc) models, which can describe the nonlinear dynamic behavior of aircraft modifications or perform incremental flight-testing.

R.5) Provide adequate resources to future programs for more incremental flight-testing when large configuration changes significantly deviate from the initial design concept.

R.6) Consider implementing mitigations or hardware systems for returning a vehicle back into a safe flight envelope without catastrophic results when performing hazardous or envelope expansion testing but insure this does not increase overall risk of such testing.

R.7) Safety experts should explore the possibility of developing improved and believable criteria for more rigorous assessment of the probability of occurrence of identified hazards for one of a kind research aircraft.

R.8) Develop a human/vehicle system interface to better conduct research for this class of vehicle.

R.9) Improve the pilot/crew displays to allow better recognition and situational awareness of slow developing hazardous events.

R.10) Consider adding attitude indicator to improve pilot's situational awareness.

R.11) Develop a method to measure wing dihedral in real-time with a visual display available to the test crew.

R.12) Develop manual and/or automatic techniques to control wing dihedral in flight.

R.13) Re-evaluate providing capability that allows pilot to better mitigate unusual flight occurrences.

R.14) Further refine the roles and responsibilities of the crewmembers to improve overall team response to unexpected and anticipated emergency conditions.

R.15) Refine emergency recognition criteria to improve team emergency response.

R.16) Perform simulations to develop recognition criteria that identify the vehicle's response prior to and during instabilities.

R.17) Improve the fidelity of aircrew simulations to mitigate the risks associated with takeoff and landing.

R.18) Enhance the depth and independence of technical participation in the research areas of this class of vehicle.

R.19) Develop capability to perform simulations of the vehicle's response to disturbances.

R.20) Improve training program and use simulations to enhance crew resource management during normal, emergency, and unstable flight conditions.

R.21) Apply crew resource management techniques to enhance crew ability for identifying and responding to emergency and unstable flight conditions.

R.22) Improve interfaces to alleviate pilot task-saturation.

R.23) Re-evaluate pilot and test team responsibilities to optimize task management.

R.24) Extend test team responsibilities to include more participatory tasks with provisions for providing advisory status of systems operation.

R.25) Consider applying advanced atmospheric models that better predict conditions hazardous to this class of vehicle.

R.26) If appropriate, re-evaluate the need to validate advanced atmospheric models using existing and, as necessary, special observations.

R.27) Ensure that meteorological hazards in the area of operations are identified and understood.

R.28) If photo helicopters are available, consider providing capability for direct voice communication between the helicopter and the Helios pilot as long as this communication does not appreciably add to Helios pilot workload.

R.29) Consider the chase plane concept of operations to improve overall test team management of unexpected and anticipated emergency or unstable conditions.

Section 11
Lessons Learned Summary

The lessons learned from the investigation are summarized in Table 11.1 below, but separated into three categories: Technical, Risk Assessment, and Training.

Table 11.1 – Lessons Learned

Technical

LL.1) Including large point masses on this type of airframe should not be attempted without optimizing the design of the primary load carry structure.

LL.2) Measurement of wing dihedral in real-time should be accomplished with a visual display of results available to the test crew during flight.

LL.3) Procedure to control wing dihedral in flight is necessary for the Helios class of vehicle.

LL.4) Time domain design and analysis tools for examining the effects of disturbances on the behavior of highly flexible vehicles are required.

LL.5) Model fidelity and validation, as well as time domain simulation, can significantly reduce technical risk where the complexity and nonlinearity of subsystem interaction is significant.

LL.6) Using numerical simulation models it is possible at modest cost to gain useful meteorological information that highlights the regional weather peculiarities to assist in preparing for flight-testing.

LL.7) Design and analysis tools applicable to large, lightweight flexible wing aircraft require better space-time domain models of atmospheric disturbances.

Risk Assessment

LL.8) Current methodology of applying engineering judgment as the sole determiner to hazard categorization for research technologies more than likely underestimates the probability of occurrence of associated hazards.

LL.9) Programs that result in vehicle usage being significantly extended beyond the initial design requirements may increase project risk more than anticipated.

LL.10) Effective risk minimization of technical risk associated with leading edge technologies may require in-depth participation by an independent technical group.

Training

LL.11) Flight crew training syllabus development could benefit by cross talk and visits to UAV developers and operators.

LL.12) Lack of crew resource management techniques and methodologies can significantly handicap a test team's ability to successfully negotiate unanticipated emergency or unstable conditions.

Acronyms

AFRL	Air Force Research Laboratory
AFSRB	Airworthiness and Flight Safety Review Board
ARC	Ames Research Center
AV	AeroVironment, Inc.
CDR	Critical Design Review
CL	Lift Coefficient
DE	Dynamics Engineer
DFRC	Dryden Flight Research Center
DOD	Department of Defense
DRR	Deployment Readiness Review
EAS	Equivalent Airspeed
ERAST	Environmental Research Aircraft and Sensor Technology
FCS	Flight Control System
FE	Finite Element
GCS	Ground Control Station
GPS	Global Positioning System
GRC	Glenn Research Center
HALE	High Altitude-Long Endurance
HAULE	High Altitude Ultra-Long Endurance
HP99	Helios Prototype Flown in 1999
HP01	Helios Prototype Flown in 2001 (also HP100k)
HP03	Helios Prototype Flown in 2003 (also H03)
HP03-1	Helios Prototype Flown 7 June 2003 (also HFC03-1)
HP03-2	Helios Prototype Flown on 26 June 2003 (also HFC03-2 and H03-2)
IO	Investigation Organizer
IWG	Independent Working Group
IRT	Independent Review Team
Iyy	Pitch Inertia
JSRA	Joint Sponsored Research Agreement
k	1000
KSC	Kennedy Space Center
LaRC	Langley Research Center
LOL	Loss of Link
MFD	Mobile Flight Director
MHPCC	Maui High Performance Computing Center
MIB	Mishap Investigation Board
MOA	Memorandum of Agreement
MP	Mobile Pilot
MIB	Mishap Investigation Board
MSR	Mission Success Review
NASA	National Aeronautics and Space Administration
NCAR	National Center for Atmospheric Research
NOAA	National Oceanic and Atmospheric Administration
PE	Planning Engineer

PDM	Project Data Memo
PDR	Preliminary Design Review
PFCS	Primary Fuel Cell System
PMRF	Pacific Missile Range Facility
RF	Radio Frequency
RFCS	Regenerative Fuel Cell System
S&C	Stability and Control
S&CE	Stability and Control Engineer
SAS	Stability Augmentation System
SP	Stationary Pilot
UAV	Uninhabited Aerial Vehicle
UH	University of Hawaii

Units

deg/sec	degrees per second
ft	feet
ft/min	feet per minute
ft/sec	feet per second (also fps)
hp	horse power
hz	hertz
kts	knots
km	kilometers
kW	kilo watt
lbs	pounds
m/sec	meters per second
m	meters
mb	millibar
psf	pounds per square foot
sec	second

Mishap Investigation Report Endorsement

Report Title: Helios Mishap Investigation Report
Date of Report: January 2004

Recommend Approval	Recommend Rejection	Endorsement and Comments
X		The Helios Mishap Investigation Report has been prepared as directed by the appointment letter and meets the requirements specified in NPR 8621.1.
		The report adequately describes the proximate cause, root causes, and contributing factors.
		The mishap report includes sufficient facts to adequately substantiate the findings.
		The Executive summary is satisfactory with the following exceptions: • Non-concur with the last paragraph of the executive summary. This paragraph includes statements that are not discussed in the body of the report, and are not substantiated with facts. Furthermore, statements concerning "strategic implications for the nation" and whether "adequate knowledge exists to design, develop, and deploy operational HALE systems" are beyond the scope of the MIB's investigation responsibilities as described by the appointment letter.
		The report provides recommendations that track with the findings and are adequate with the following exceptions: • Non-concur on the following recommendations: R3, R9, R17, R21, and R24. Rationale: The recommendations either provide a solution for a problem that was not described in the report or in general describe "good practices" and not specific solutions. The appointing official may address these recommendations in the Corrective Action Plan (CAP) at his discretion.
		Concur with the intent of the following recommendations but believe that as written, they will be difficult to implement because they are vague, unverifiable, or unachievable: R5, R6, R7, R8, R9, R13, R14, R18, R20, R22, R23, R26, R27, R28, R29. Recommend that the appointing official ensure that the CAP translates these recommendations into clear, verifiable, and achievable actions that adequately address the relevant findings.

NASA Endorsing Official

Signature	Date of Endorsement
Bryan O'Connor (signature)	12 APRIL 04
Printed Name Bryan O'Connor	**Title** Associate Administrator Office of Safety and Mission Assurance

Mishap Investigation Report Endorsement

Report Title: Helios Mishap Investigation Report
Date of Report: January 2004

Approved	Recommend Rejection	Endorsement and Comments
✔		The Helios Mishap Investigation Report has been prepared as directed by the appointment letter, meets the requirements specified in NASA Procedures and Guidelines (NPG) 8621.1, and was reviewed in accordance with NPR 8621.1. The report clearly describes the proximate cause, root causes, and contributing factors to the Helios mishap. Factual evidence and supporting analyses are provided to substantiate each of the findings presented in the report. Recommendations provided by the investigation team are appropriate, and offer a sound basis to strengthen risk management of future efforts on high-altitude, long-endurance aircraft technology.

NASA Endorsing Official

Signature	Date of Endorsement
Lebacqz 5/13/04 Printed Name Title J. Victor Lebacqz Associate Administrator Aeronautics Research Mission Directorate	**May 13, 2004**